PROJECT MANAGEMENT FOR IT-RELATED PROJECTS

BCS, THE CHARTERED INSTITUTE FOR IT

Our mission as BCS, The Chartered Institute for IT, is to enable the information society. We promote wider social and economic progress through the advancement of information technology science and practice. We bring together industry, academics, practitioners and government to share knowledge, promote new thinking, inform the design of new curricula, shape public policy and inform the public.

Our vision is to be a world-class organisation for IT. Our 70,000 strong membership includes practitioners, businesses, academics and students in the UK and internationally. We deliver a range of professional development tools for practitioners and employees. A leading IT qualification body, we offer a range of widely recognised qualifications.

Further Information
BCS, The Chartered Institute for IT,
First Floor, Block D,
North Star House, North Star Avenue,
Swindon, SN2 1FA, United Kingdom.
T +44 (0) 1793 417 424
F +44 (0) 1793 417 444
www.bcs.org/contact

PROJECT MANAGEMENT FOR IT-RELATED PROJECTS
Second edition

Bob Hughes (editor)
Roger Ireland, Brian West, Norman Smith,
David I. Shepherd

bcs
The Chartered Institute for IT

Published by BCS Learning & Development Ltd, a wholly owned subsidiary of BCS The Chartered Institute for IT, First Floor, Block D, North Star House, North Star Avenue, Swindon, SN2 1FA, UK.
www.bcs.org

ISBN: 978-1-78017-118-0
PDF ISBN: 978-1-78017-119-7
ePUB ISBN: 978-1-78017-120-3
Kindle ISBN: 978-1-78017-121-0

British Cataloguing in Publication Data.
A CIP catalogue record for this book is available at the British Library.

Disclaimer:
The views expressed in this book are of the author(s) and do not necessarily reflect the views of the Institute or BCS Learning & Development Ltd except where explicitly stated as such. Although every care has been taken by the authors and BCS Learning & Development Ltd in the preparation of the publication, no warranty is given by the authors or BCS Learning & Development Ltd as publisher as to the accuracy or completeness of the information contained within it and neither the authors nor BCS Learning & Development Ltd shall be responsible or liable for any loss or damage whatsoever arising by virtue of such information or any instructions or advice contained within this publication or by any of the aforementioned.

Typeset by Lapiz Digital Services, Chennai, India.
Printed and bound by CPI Group (UK) Ltd, Croydon, CR0 4YY.

CONTENTS

FIGURES AND TABLES

PREFACE

This book is designed to give a practical introduction to IT project management principles and techniques. The first edition was intended more specifically to support candidates for the BCS Foundation Certificate in IS Project Management This edition still supports this qualification, but updates some of the material and broadens its practical application.

Taking this qualification is not itself a daunting challenge: it consists of an hour-long 40-question multiple choice examination. However, the intention was never just to help cram for an examination. While there might be an immediate concern to pass a test, for most people the more important motivation was to gain guidance on planning and managing an IT project. The text was designed to help those from an IT practitioner background who were beginning to take on project management responsibilities. However, it is not just IT developers who have to grapple with IT projects: users often have to bear the brunt of IT-driven business change and have their own project responsibilities that can have a decisive impact on project success. An additional aim was to give these IT users some insights into IT project management issues. The text therefore goes beyond simply helping people to tick the right boxes in a test and aspires to support novice IT project leaders in their place of work.

When learning about any new topic, a good starting point is a text which provides a simple explanation of the basics. This can provide a foundation that allows you to go on and grasp more advanced concepts. A measure of the success of the first edition was that it started to be used for purposes for which it was not primarily designed. The text started to appear in the reading lists of courses where the overall syllabus was broader and the assessment more demanding than the foundation certificate. One example of this was the BCS Higher Education Qualification Diploma in Project Management (an 'academic' BCS qualification comparable to a UK university award and taken mainly by overseas candidates). Some of the changes for the second edition have been made in response to this unexpected use. The focus still remains on the foundations but care has been taken to provide links to other, more detailed project management material. Wherever possible, alternatives to the terminology we have used are provided for techniques and concepts to allow easier cross-reference to other bodies of knowledge. For example, 'steering committee', 'project board' and 'project management board' all refer to largely the same concept in project management.

We have put in links to further material using a 📖 symbol for those who want to explore a topic more deeply. Some material in the basic text goes beyond what is needed for the BCS foundation syllabus and these have been marked with a 🎓 symbol to indicate an 'advanced topic'.

It may be heretical to say this in a project management text, but successful projects do not depend only on good project management and some of the links provided are to material on complementary disciplines that can assist positive project outcomes. (The BCS Diploma in Business Analysis to which the Foundation Certificate in IS Project Management can contribute supports this view.)

The BCS Foundation Diploma syllabus has been very stable in recent years and there have been no massive changes in content in the new edition. Some inadvertent gaps have been filled: for example, more has been added on the question of deciding whether to build or buy an IT solution. A suggestion to acknowledge the growing interest in agile approaches has been incorporated. The main principles of project agility – such as the focus on iterations and increments in project delivery – had been well-established before the term 'agile' was adopted in the context of software development, so it has been easy to signpost those elements of our approach that dealt with them. It was also suggested that more quantitative approaches to risk assessment be discussed, and this has been done.

The BCS Foundation Certificate course's focus is different from that of PRINCE2. PRINCE2 is a UK government-sponsored set of procedures for managing major projects. In our view, it effectively describes an information system for a project that allows it to be run in a controlled and efficient manner. Although PRINCE2 is really an administrative standard that will tell you what decisions need to be taken, when they need to be taken and by whom, it offers little guidance about how decisions are made: it does not claim to be a set of project management principles and techniques.

The BCS syllabus can also be distinguished from more general introductory courses on project management by its focus on IT projects. While the core elements of project management remain the same regardless of the type of project, there are some significant differences in emphasis with IT projects. The description of the IT-focused system development life cycle has already been mentioned, but there are other topics – like testing and the measurement of functionality to support the estimation of system size – which get more attention here than in more general project management courses.

The following people contributed the material for the text:

Norman Smith Chapters 1 and 4
Bob Hughes Chapters 2 and 6
Roger Ireland Chapters 3 and part of 8
Brian West Chapters 5 and part of 8
David I. Shepherd Chapter 7

Any defects and errors are almost certainly those of the editor, Bob Hughes. Sue McNaughton and Elaine Boyes at the BCS drove the publication project for the first edition along. The original development of the Foundation Certificate as a whole has involved many BCS staff, including Malcolm Sillars, Rebecca Stoddart, Imelda Byrne, Steve Causer and Carol Lewis. Jutta Mackwell was instrumental in the creation of this second edition. Roger Ireland has been a painstaking reviewer, and Karen Greening managed the production of the book from the author's word-processed manuscript to the final version that appears here.

The book is dedicated, as was the last, to the memory of Jimmy Robertson.

USEFUL URLS

IT Project management qualifications and syllabuses

- BCS Professional certification: Foundation Certificate in IS Project Management http://certifications.bcs.org/content/ConTab/2

- BCS Professional certification: Foundation Certificate: Programme and Project Support Office Essentials http://certifications.bcs.org/content/ConTab/3

- BCS Higher Education Qualifications: Diploma in IT Project management. An 'academic' examination at university 2nd Year level popular with overseas candidates. www.bcs.org/upload/pdf/dippmsyll.pdf

Agile project management approaches

- DSDM Consortium. This group is responsible for the DSDM Atern agile project management framework www.dsdm.org/

- Scrum Alliance has a set of resources supporting the Scrum agile framework. www.scrumalliance.org/

- **Some professional bodies** – APM and PMI have their own qualifications

- PROMSG: the BCS Project Management Specialist Group www.proms-g.bcs.org

- Association for Project Management: the UK professional body for generic project management (rather than just IT) www.apm.org.uk

- Project Management Institute: US-based professional body www.pmi.org. A UK chapter of PMI exists www.pmi.org.uk

- International Project Management Association. A global umbrella association to which most national project management professional bodies are affiliated http://ipma.ch

Planning tools

- Microsoft Project : probably the most widely used project planning tool www.microsoft.com/project/en-us/project-management.aspx

- Oracle Primavera: another, perhaps more industrial, project planning tool (and much else) www.oracle.com/eppm

- Smartsheet is an easy to use tool for small one-off projects where there is a need to do things quick and simple way

Quality

- TickIT: UK initiative to tailor ISO9001 to specifically IT development. www.tickit.org
- Details of the SEI Capability Maturity Model (CMMI). www.sei.cmu.edu/cmmi/

Estimation and measurement

- UK Software Metrics Association contains useful information and/or links relating to function points of various types. www.uksma.co.uk/

Project organisation

- PRINCE2, the UK government- sponsored standard for project management procedures. www.prince-officialsite.com/

General keeping up to date

- Project Management Today, trade magazine www.pmtoday.co.uk/

1 PROJECTS AND PROJECT WORK

LEARNING OUTCOMES

When you have completed this chapter you should be able to demonstrate an understanding of the following:

- *the definition of a project;*
- *the purpose of project planning and control;*
- *the typical activities in a system development life cycle;*
- *system and project life cycles;*
- *variations on the conventional project life cycle;*
- *implementation strategies;*
- *the purpose and content of the business case;*
- *types of planning documents;*
- *post-implementation reviews.*

1.1 PROJECTS

A **project** may be defined as a **group of related activities carried out to achieve a specific objective**. Examples of projects include building a bridge, making a film and re-organising a company. We will be focusing on projects that implement new information technology (IT) applications within organisations. These are technical but also involve changing the organisation in some way.

Before a project starts, one or more people will have an idea about a desirable product or change. Before this idea can become the subject of a project, a **business case** will need to be made showing that the value of the benefits of completing the project will be greater than the costs of implementing and operating the new (or revised) system that the project would create. This will not only need to consider business concerns, but also the technical difficulties of the project. This is underlined by the alternative name of **feasibility study** for the business case.

COMPLEMENTARY READING

Exploiting IT for Business Benefit, Bob Hughes, BCS

If the proposal is accepted by the organisation, the project that emerges should have the following attributes:

- A defined start point, which is when:
  - the exploration of the idea is converted into an organised undertaking;
  - the idea obtains business backing and a **project sponsor** – an individual or group within the organisation who will take ownership of the project and ensure that it has the appropriate financial resources;
  - a commitment is made to provide the necessary resources;
  - responsibilities are defined;
  - initial plans are produced.
- A set of objectives, which:
  - drive the actions of the project team towards achieving a common goal;
  - should be stated and understood at the start of the project;
  - should be clear and unambiguous.
- A set of outputs or deliverables, which allow the objectives to be satisfied.
- A date by which the objectives should be met and a budget setting the maximum allowable cost of the project.
- A unique purpose – routine activities are not projects.
- Benefits for the organisation which justify carrying out the project, and which are:
  - ideally, measurable;
  - greater than the costs.

In some cases, the cost of the project might be greater than the immediate benefits, but completion of the project may enable other projects to be implemented which will reap the benefits – this is often the case with IT infrastructure projects.

1.2 SUCCESSFUL PROJECTS

To be successful, a project should:

- enable the stated objectives to be achieved;
- be delivered on time and within budget;

- deliver a system that performs to agreed specifications, including those relating to quality;

- satisfy the project sponsor and other interested parties. The term **stakeholder** refers to anyone who has an interest in the project: their role is discussed further in Section 8.3.

These are **project** objectives. The IT functions that are delivered, including both hardware and software, should enable the organisation to meet its **business** objectives. For example, the development of a new website could enable an organisation to sell its products online to a wider market. However, while the project objective of delivering the new website might be achieved, the business objective of selling more products might be denied because of external factors such as a general downturn in the market.

Objectives are often called **success criteria**. If they are satisfied then the project can be deemed a success. They should focus on the desired state of affairs that should exist when the project is completed, rather than on the details of how the project is to be done.

There is a mnemonic to help recall the characteristics of good success criteria. They should be SMART (**S**pecific, **M**easurable, **A**chievable, **R**esource-constrained and **T**ime-constrained). The success criteria should be specific and measurable – for example, 'increasing market share' is too vague: this could be the outcome of lots of different activities and the amount of increase expected is not defined. 'Creating an online booking system that will be used by at least 30 per cent of customers in its first year' is more specific. As well as being specific and measurable, success criteria must be clearly achievable. If it is clear that they are not, people are likely to ignore them. Finally, part of the statement of objectives will always define a deadline and an overall target cost.

In summary, this means developing the project at a **specified cost**, within a **specified time**, to meet a **specified business requirement**. These three specifications are closely linked and any change to one will affect the others. The project objectives which relate to cost, time and the degree to which requirements are satisfied (**'scope'**) are often called the **'iron triangle'**.

The sponsor and users typically want a system with a broad scope – capable of a multitude of functions – to be delivered immediately and at low cost. As a general rule, not all of this can be delivered, and so the agreed project objectives will be a compromise between the three corners of the iron triangle of cost, time and scope.

If the scope of requirements strays outside this area of compromise, it will increase cost or delivery time and the project could cease to be viable. The costs of the project could as a result exceed the value of the benefits of the project. Generally an increase or decrease in any of the three factors of the triangle will affect the others. Thus if the deadline for project completion has to be brought forward, either the scope would have to be reduced or more staff could be employed to work in parallel on the project, which would increase costs – and some project risks. There may be exceptional circumstances in which a project can, with the sponsor's agreement, fail to meet one or more of these success criteria and yet still be considered a success, usually because the business objectives can still be met.

While this book-keeping element of successful project management is important, note that the perceived value of the benefits of a project may be quite subjective. The final judges of the success or failure of a project will be the project sponsor and the users of the delivered IT applications. Being sensitive to their needs will be as important as sticking to the letter of a contract.

CANAL DREAMS BOOKING SYSTEM PROJECT SCENARIO

Canal Dreams is a major holiday company that specialises in canal holidays. The present organisation is the result of the acquisition of six regional canal boat leasing operators, as Canal Dreams has expanded over a number of years. Currently there is a central call centre that deals with holiday bookings using an in-house computer-based booking application, which is now some years old.

Business development analysts have identified the need for an enhancement of the system so that customers can book boating holidays directly over the internet. One advantage of this is a possible increase in bookings by overseas customers.

The current IT system was developed by an external software development company some time ago, and currently there are two software developers who maintain the current system and implement relatively minor enhancements. However, the management of Canal Dreams does not believe that it has sufficient in-house resources to develop the new functionality, particularly as they see the required extension to the system as an urgent business need. The intention is therefore to contract out the system design and building of the system to an external company.

This Canal Dreams project scenario will provide examples throughout the text. The main objective in this scenario is the enhancement of the Canal Dreams holiday booking system to allow potential customers to book holidays over the web. This will have business benefits for Canal Dreams. With the new system, potential customers can browse an online brochure of boats and start and finish locations, check if there is one available where and when they wish to go on holiday and, if available, make the booking – all via the internet. The new system will thus make booking possible 24 hours a day and seven days a week, and this improved accessibility, it is hoped, will increase sales. An automated online system should eventually allow staff reductions as the internet becomes the preferred medium for bookings.

In order to meet these business objectives, the proposed system will need certain functionalities. For instance, it should allow the potential customer to check the availability of a boat at a particular boatyard in a particular week. These **functional requirements** will include not just those of the organisation but also **legal requirements**, such as those relating to distance selling.

There will also be **quality requirements**: for example, the time it takes the computer to respond to a user query on boat availability needs to be quick so that frustrated potential customers do not abandon their queries and try a competitor's website.

If the system were to exceed the **cost requirement**, the potential additional income through extra bookings and staff savings might not be enough to meet the cost of implementing the system. Canal holidays are a seasonal business and so implementation of system enhancement will need to be at a quiet time of the year before the bookings for the next season start to come in. This implies a certain **deadline** for system implementation.

1.3 PROJECT MANAGEMENT

Having established and agreed objectives, how do we then achieve them? The first step is good planning. Having produced good plans, monitoring and effective control of the project is needed to fulfil the plans and achieve the agreed objectives. Someone has to take responsibility for controlling the work in accordance with the plans. This is the role of the **project manager**.

A successful project cannot be guaranteed, but certain things will contribute to success, such as:

- **Clearly defined responsibilities:** it is essential that project roles and responsibilities be clearly defined, documented and agreed.

- **Clear objectives and scope:** any manager who embarks upon a project without clearly establishing the scope of the expected deliverables, together with cost, time and quality objectives, is creating problems for the future. These should be laid down in **terms of reference** or some other document that defines the scope of the project objectives.

- **Control:** despite their individual differences, all projects can be controlled. It is important to establish at the outset how best to control the work and how to exercise that control.

- **Change procedures:** ideally the project manager would like to work in a world where there is no change or uncertainty. Unfortunately it has to be recognised that there is uncertainty and that change will happen. Appropriate change procedures are needed to deal with it.

- **Reporting and communication:** clear reporting of project progress and any problems allows action to be taken quickly to resolve problems. Effective communication with all stakeholders can help avoid conflicts.

A **project management method** is a set of processes used to run a project in a controlled and, therefore, predictable fashion. The design and development procedures by which the objectives of the project are satisfied – for example, the use of object-oriented analysis – constitute the **development methods**.

There are various project management methods which complement the development methods that can be used – a well-known one in the UK is PRINCE2. In general, project management methods are applicable to a range of project types, whereas development methods tend to be specific to projects with particular types of deliverables or objectives. This is because the development tasks will vary according to the objectives of the project. Organising an office move, developing a software application and providing disaster recovery facilities are all projects in their own right. Each has a different method of development but all are controllable using the same project management processes. Following a method does not guarantee that a project will be successful. If applied carefully, however, it will provide management with the means to be successful.

ACTIVITY 1.1

Assume that you are a manager of an office department that is going to be relocated to a building five miles away. Day-to-day management of the move will be delegated to one of your staff. What would be the main sequence of activities needed to plan and carry out the move? You want to leave as much of the detailed work as possible to your subordinates, but at which key points would you need to be involved to check progress?

Solution pointers for the activities can be found at the end of the chapter.

1.4 SYSTEM DEVELOPMENT LIFE CYCLE

Dividing a development method into a number of processes is a widely accepted practice. This allows systems to be designed and implemented using a methodical and logical approach. The number and names of these processes will vary from organisation to organisation. In some cases, stages will be combined or split. Generally speaking, the following processes belong to the **system development life cycle** (SDLC) that applies to IT projects:

- initiation;
- identification of the business case;
- project set-up;
- requirements elicitation and analysis;
- design;
- construction;
- acceptance testing;

- implementation/installation;
- review and maintenance.

This suggests a particular sequence of processes. However, different parts of an application under development could be at different stages. For example, one component could be still being designed while another is being coded. The key point is that all these technical processes have to be dealt with somewhere within the project.

Each process creates one or more tangible products or **deliverables**. Delivery of the products of each process can act as a **milestone** at which we can judge the progress and continuing viability of the project.

One variation of this model is where software and/or hardware components are to be acquired off the shelf. Because the components already exist, the design and construction processes are not carried out. Instead, a selection process is devised consisting of methods of evaluating the suitability of candidate products. The products to be used are then selected. Some element of customisation may be needed to modify the product for use in the organisation. An acceptance test could be carried out.

ADVANCED TOPIC Build versus buy

The question of whether in fact new software has to be written must always be asked. Existing software bought off the shelf has the following advantages:

- It already exists, so it can be installed more quickly.
- It can be seen in action, so users can get a good idea of its quality.
- Existing users will have effectively tested it by reporting any defects, which will then have been removed by the supplier, so the software is likely to be more reliable.
- As there are many different organisations using the software, development costs will be shared and the cost of the software should be cheaper than if you had built it yourself.
- You do not have to employ software developers to build the new system, who may then become surplus to requirements.
- The vendor should supply updates to deal with statutory changes so maintenance of the installed system will not be a responsibility of the host organisation.

However, there are disadvantages in buying off-the-shelf software, which may encourage the building of new software:

- Off-the-shelf software may not meet all the particular requirements of the host organisation.
- The organisation may have to change its business processes to fit in with the way in which the off-the-shelf application works.
- If you adopt an off-the-shelf application, you can be as good as your competitors – who may have the same system – but you cannot be better than them.

- Once you have adopted a particular off-the-shelf package, it may be difficult to change. Off-the-shelf software is often leased on an annual licence and if the vendor increases the licence fee, you may be trapped into having to pay it.

- If the vendor ceases to trade, this may put you in a difficult situation if changes to software are needed.

1.4.1 Initiation

The first two processes – initiation and the identification of the business case – are, strictly speaking, not part of the project. Their purpose is to establish the justification for the project: an outcome could be a decision not to go ahead with the project, which means the planning of the project only starts in earnest after these two processes have been completed.

The objective of project initiation is to decide the most appropriate way to respond to a request for some work to be done, taking into account any business or technical strategies that the host organisation might have.

It begins with recognition by an organisation's managers that it has a need that can only be satisfied by some form of project. The need might be a perceived problem to be solved, a request for something new to add to an existing system or the identification of some new way of delivering value to the organisation. The initiation process checks that a problem or opportunity really exists and decides whether the proposed change appears to be desirable, and whether a project is the best way to implement the change. This phase is typically short. The end result is a decision by the project sponsor on whether resources should be spent on further investigation of the feasibility of the proposal, including the business case for it. **Terms of reference** should be drawn up, outlining the scope of the proposal to be investigated and authorising staff to carry out the investigation. Staff carrying out the investigation need to have permission to gather information from those working in the areas affected, along with other stakeholders.

1.4.2 Identification of the business case

The business case or feasibility study assesses whether the proposed development is practical in terms of the balance of costs and benefits, the technical requirements and the organisation's information system objectives. The deliverable is a feasibility report which, among other things, presents the business with a range of options aimed at providing a solution.

Apart from the question of business viability – the benefits being greater than the costs – factors influencing the decision about the appropriate option include:

- **Budget constraints:** the benefits may be greater than the estimated costs, but does the organisation have the resources to pay for the investment?

- **Technical constraints:** can the proposed project be completed with the technology currently available?

- **Time constraints:** can the proposed project be completed in the available time?

- **Organisational constraints:** can the organisation cope with the changes that the new development will demand?

One or more of these constraints may prevent a project from being developed any further. The content of the business case is discussed in more detail in Section 1.9.

1.4.3 Project set-up

Based on the recommendation of the **business case report**, the organisation will decide whether to go ahead with the full project. At this point a group variously called the **steering committee**, **project board** or **project management board** is set up to oversee the project in the organisation's interests. A project manager needs to be appointed and an initial project team set up to start work.

More detailed planning for the project takes place. Important decisions will be taken about how the declared project objectives are to be fulfilled. **Terms of reference** for the project to implement the new system, as opposed to simply investigating its feasibility, are drawn up.

1.4.4 Requirements elicitation and analysis

This phase defines the requirements of the new system in detail and identifies each business transaction. Some work on identifying requirements will already have been done when the business case was being identified. The elicitation or gathering of requirements could involve:

- interviewing users and their managers;

- examining documentation describing the current operations;

- analysing operational records created by the current system;

- observation of work practices;

- **joint application development** (JAD) sessions – where groups of stakeholders and business analysts meet in intensive (usually day-long) sessions to identify and agree detailed requirements;

- questionnaire surveys.

In some cases, mock-ups or **prototypes** of parts of the new system could be used to help the users clarify their ideas about the requirements.

ACTIVITY 1.2

What kinds of people should the business analyst interview in the Canal Dreams ebooking enhancement project in order to obtain the requirements for the new system?

Business analysis techniques such as business process modelling or data analysis will usually be applied to organise the raw data collected.

COMPLEMENTARY READING

Business Analysis 2nd Edition, edited by Debra Paul, Don Yeates and James Cadle, BCS

At the end of this phase, some form of **requirements statement** is produced. This describes what the final system should be able to accomplish and lists all the major features of the end product. It forms the basis of the contract between the customer for the new system and the developers.

At this point an outline of the test cases, consisting of test transactions and the results expected from them should be drafted. As will be seen, these will be used to check that the delivered system conforms to the requirements statement when acceptance testing is done.

1.4.5 Design

If it is decided to build a new system, rather than buying a ready-made or off-the-shelf application, then a design phase is needed. This activity translates the business specification for the automated parts of the system into a design specification of the computer processes and data stores that will be needed.

Where a new application is to be built, the elements to be designed include:

- inputs;
- outputs;
- processing;
- data and information structures.

The identification of the inputs, outputs, business rules and information that the system will process is known as **logical design**. The **physical design** is concerned with the actual appearance of the input and output screens and the printed reports that will be produced by the implemented system. Several different physical designs could satisfy the same underlying logical design. Physical design in this phase is essentially concerned with the system as it will appear to the outside world. Further **internal physical design** of the software and data structures will take place in the next phase.

Where an off-the-shelf application is to be used, because it already exists, its design will already have been carried out. In this case, the issue is finding the package whose features most closely match the business requirements. The process by which available packages are to be evaluated, and those which represent good value selected, must be planned. Evaluation may involve trying out demonstration versions of the software, site visits to existing users of the software, and careful study of the suppliers' documentation. In some cases the existing software will need to be customised – that is, modified – to meet the organisation's particular needs.

ACTIVITY 1.3

List some of the screens and other possible inputs and outputs that the new Canal Dreams ebooking enhancement might need and which might need to be designed.

1.4.6 Construction

This process has the objective of designing, coding and testing software and ensuring effective integration between different software components. For example, the new Canal Dreams holiday ebooking enhancement would need the development of a application to record holiday bookings made by customers online. It is very rare these days for new IT applications not to be linked in some way with existing applications. In the case of the Canal Dreams ebooking enhancement, an existing database of holiday bookings used by telesales staff can be accessed and updated, and much existing functionality can be used. It is therefore likely that as well as new code being developed, some existing code may need to be amended to deal with the enhancement.

Procedure manuals will also be produced and new hardware may have to be acquired. In the case of Canal Dreams, additional servers will be needed to deal with the increase in internet transactions. During this phase, the requirements statement will be re-examined to ensure that it is being followed to the letter. Normally any deviations have to be approved through a formal change procedure (see Chapter 4).

1.4.7 Acceptance testing

In Section 1.4.4, we suggested that test cases to ensure that the requirements had been met should be outlined at the requirements analysis stage. These can now be used to check the delivered system. This testing could be carried out by knowledgeable representatives of the users and IT support staff before its implementation as an operational system. It is inevitable that during this stage, the user will uncover problems that the developer has been unable to detect.

In the case of the Canal Dreams ebooking enhancement, a problem is that users of the new facility will be members of the public who we do not currently know. Usability tests using external members of the public recruited for this purpose will be needed to test the customer-facing interfaces.

These acceptance-testing activities may overlap with implementation. For example, some tests need to be carried out on the system when it is actually installed on the equipment that will be used operationally.

1.4.8 Implementation/installation

Here the project reaches fruition. Hardware that has been purchased is delivered and installed. Software is installed, users trained, and the initial content of databases set up. In the case of the Canal Dreams ebooking enhancement, the general public will need to be informed of the new facility for online booking. There are various strategies for implementation; these are discussed later in Section 1.10.

1.4.9 Project closure

Although we have put this after implementation/installation, a project could have been abandoned at an earlier stage because its business case was no longer valid.

In the case of successful project completion, certain tasks will need to be done on closure, including:

- sign-off of acceptance documents by the project sponsor – this may be conditional on other key stakeholders giving approval first;

- handing over responsibility for maintenance and support to a permanent team;

- closing down accounts relating to the project;

- the project manager writing a **lessons learnt report**;

- releasing and re-allocating project resources, including the project team and the project manager;

- arranging publicity to tell the outside world about the project's success.

1.4.9 Review and maintenance

At an agreed interval after the system has been made operational, a **post-implementation review** should be carried out by a business analyst who was not involved in the original project. The review checks that the operational system has actually delivered the benefits envisaged in the original feasibility report. Changes may sometimes result from this review if the system does not completely fulfil its original requirements, but they may also be the result of users identifying new requirements. Changes may also come about because of changes to government regulations or alterations in the way the organisation does business. These changes can be made as part of maintenance work, or they can become projects in their own right.

1.5 PROJECT MANAGEMENT AND THE DEVELOPMENT LIFE CYCLE

The processes described above represent groups of development activities which have to be performed to complete the project. There is also a need for management reviews at which the progress and direction of the project can be formally assessed.

Most projects contain elements of uncertainty which make it difficult to meet exactly all planned targets. This uncertainty tends to be greatest at the beginning of the project, when little may be known about detailed requirements or any novel technologies to be deployed, but gradually decreases as the project progresses. This makes it difficult, at the beginning of the project, to plan its later phases in detail. For example, it would be difficult for Canal Dreams to plan in detail the software code to be developed without a detailed understanding of the new requirements and how they will affect the current system.

For the purposes of control there is a need to break the project into manageable units of work. These are variously called **stages** or **phases**. This is similar to dividing development into processes, except that it focuses on how best to manage the project. No two projects will be broken down identically because no single project structure can provide the best management control for all projects. In some cases – perhaps in smaller projects – two activities of the system development life cycle, such as design and construction, might be treated as one stage for management purposes. On larger projects a single activity of the development life cycle might be split into several management stages. For example, the construction phase might be broken into different management stages dealing with the construction of different parts of the system.

The end of each stage is marked by a formal review, involving the project sponsor, which assesses the work done and the work still to be done on the project as a whole, and whether the business case is still valid. The review concludes with a formal sign-off of the last stage and the project sponsor's approval to continue to the next stage of the project.

1.6 ELEMENTS OF PROJECT MANAGEMENT

The motivation behind consolidating or splitting up development processes is to ensure controllability of work. In addition, the project manager has to tailor project management procedures to maintain control over the project, while avoiding excessive management bureaucracy. Although the overall project management process remains the same, different projects require different levels of control. The amount of effort required for project management, for example, needs to be appropriate to the size of the project. However, the fact that a project is small does not mean it requires no control.

The processes that need to be tailored relate to the following:

- planning and estimating;
- monitoring and control;
- issue management;
- change control;
- risk management;
- project assurance;
- project organisation;
- business change management.

All of these will be described in greater detail in later sections of this book, but a brief overview will be useful at this point.

1.6.1 Planning and estimating

Good planning increases confidence within the project team. Without a plan there is no means of knowing whether dates can be achieved, nor whether the project is adequately resourced.

The aim is to detail all the activities, the sequence in which they are carried out and the resources they need. The plan shows when activities are to be performed and helps to estimate the staff effort needed and the most likely finish dates. An **outline** plan for the whole project will be made initially, then a **detailed** plan for each stage will be created nearer the time at which that stage is to start. Chapter 2 looks at planning in more detail.

1.6.2 Monitoring and control
The project plans form the basis for monitoring progress against the expected achievements. Tracking and control aim to ensure that the project meets its commitments in terms of deliverables, quality, time and cost by tracking the current state of the project against the plan and identifying any need to re-plan. Chapter 3 examines project control in more detail.

1.6.3 Issue management
During the course of a project, problems will be identified which are considered likely to affect the project's success. These problems or project issues may or may not be within the control of the project manager, but do not include authorised changes to the project requirements – these have their own special procedures (see Section 1.6.4). The project manager should ensure that a system is in place for recording these issues, monitoring their status and starting any actions needed.

1.6.4 Change control
Any project will be subject to change. A change may result from a modification to requirements or may come as a result of errors found in testing. Any requests for change should be made through a formal change management process. Failure to incorporate necessary changes reduces the benefit obtained from the project. However, accepting changes in an uncontrolled manner can cause problems related to the cost, time scales and overall business case for the project.

Chapter 4 examines change control and configuration management.

1.6.5 Risk management
All projects are subject to risk. If these risks are not managed, they can have a detrimental effect. A suitable risk management process assesses and manages project risks.

Risks are different from issues. A risk is an unplanned occurrence which could happen, but has not yet done so. An issue is an unplanned occurrence which has already happened and which requires the project manager to request or initiate action not previously planned.

Risk management identifies and quantifies risks before they happen, and plans and implements actions to eliminate risks or reduce their probability or impact. Risk management ensures that projects are only undertaken with a full understanding of the potential implications of the risks involved. Chapter 7 deals with risk in more detail.

1.6.6 Project assurance
As noted above, project monitoring and control involves assessing various aspects of work, including progress, cost, changes, issues, risk and quality. When pressure

mounts on a project to meet its deadline, it is tempting to ignore some of the checks and balances imposed by project control and to focus exclusively on the work to be done. This lack of control can be dangerous and can lead to project failure. Project assurance is a set of procedures which ensures correct project control is maintained. This involves auditing by staff outside the project team.

1.6.7 Project organisation
A key factor in any project is effective project organisation, wherein the roles and responsibilities of all participants are clearly defined and understood. Chapter 8 explicitly addresses this topic.

1.6.8 Maintaining stakeholder engagement
Winning stakeholders' support for a project is important for project success. Unless time has been spent communicating with stakeholders and making sure that users know exactly what to expect of the new system, the project could meet its formal requirements but still be seen as a failure by its users.

1.7 DEVELOPMENT PROCESS MODELS

It was noted in Section 1.4 that organisations involved in IT development need a well-defined, repeatable and predictable system development life cycle. Projects creating different types of products need different development life cycles. Whatever the life cycle, it will be a set of activities which, after completion, results in one or more products that are delivered to a customer. Each activity in the process will have a defined input and output.

Effective development methods have certain characteristics. They are made up of an overall set of techniques and activities from which team members working on a new project of a particular type can select the most appropriate subset. The method should never require a task that does not produce something useful to the project.

The conventional system development life cycle for IT projects was described in Section 1.4. This has certain general characteristics that could apply to a number of different life cycles. However, it assumes that **technical processes** will vary according to the type of project. For example, it distinguishes as different project types (with differences in the types of processes carried out) those in which software is to be developed and those in which existing software is to be obtained off the shelf. In Section 1.5, however, it was noted that to make projects more manageable, the processes could be split up or sequenced in different ways. Here we look at some of the options for this.

1.7.1 Waterfall model
The waterfall model (see Figure 1.1) is the basic phased model of a development cycle. It is also known as the one-shot or once-through approach. The model takes its name from the way each phase cascades into the next. It is assumed that activities are normally done in a strict sequence, although there is some scope for re-working stages once they have been completed. Phases should produce a sequence of deliverables, such as the requirements statement, design documents and software structures, where the output from each phase is an input to the next.

Figure 1.1 The waterfall model

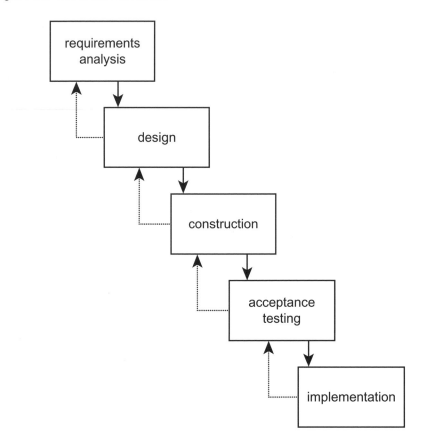

This approach provides for feedback loops which are activated when there is a need to revisit an earlier stage to redesign, recode and so forth. It is possible to return to any previous phase, although this could well require extensive replanning. The ideal is for quality control activities to be associated with each phase, so that once the deliverables of a stage have been signed off they should not need to be reworked.

This model is probably best used on projects where requirements have been clearly defined and agreed. Unfortunately, most projects do not have clear requirements at the beginning. As the model relies upon having each phase completed and signed off, it can become bureaucratic and time-consuming. It works best where there are few changes to requirements during the development cycle.

Another drawback is the amount of project documentation which can be created. The distinct testing phase at the end of the project means major defects could be undetected until late in the project, when they are more difficult to repair. It is also easy to misjudge progress: just because the requirements have been signed off, it does not necessarily mean the requirements have been clearly understood.

1.7.2 Agile project practices

ADVANCED TOPIC Agile practices

A response to the limitations of the waterfall approach has been that some have argued for the adoption of what are called **agile** practices. These tend to be practices that reduce bureaucratic obstacles by encouraging intense, informal, communication between project participants. Some of these work at the level of individual work teams. **Scrum**, for example, breaks a project into a number of increments (see Section 1.7.3) of about two-week duration called **sprints**. Within sprints, Scrum breaks individual activities into a series of small steps which are listed in a **backlog**. Each day the project team members report on their progress in implementing the items in the backlog in short 'stand-up' meetings. Other agile approaches – such as **Extreme Programming (XP)** – focus on software development practices. XP, for example, calls for software developers to work together in pairs, so the coding decisions of one developer are always checked by the other as the code is entered at a work station. Another XP practice is the integrating of all existing software components on a daily basis to ensure consistency.

From the point of view of those who manage projects, a common theme of the different agile approaches is the adoption – to differing degrees – of the principles that support the incremental and iterative models described in the next two sections, and which actually predate the first use of the description 'agile' (and are part of the BCS Foundation Certificate syllabus). The **Dynamic Systems Development Method/Atern Method (DSDM Atern)** is the approach that most closely reflects the principles described in the next two sections.

We will be returning to other relevant agile practices in later chapters, particularly Chapter 5, which discusses quality issues.

1.7.3 Incremental model

Although the incremental model (see Figure 1.2) is similar to the waterfall model, it involves the development and delivery of functionality in fragments or **increments**. Typically, global requirements are defined and an overall architecture designed. Then the product is developed in increments. After each increment is designed, developed and tested, it is system tested and then becomes operational, so that users get their new system in instalments. This approach works best when the requirements are relatively well-known. It can work well with larger projects, as these are effectively broken into a series of mini-projects, each delivering an increment.

The incremental model is often used in conjunction with **timeboxes**. The deadline for completion of the increment is fixed and the features to be delivered by the increment are ranked according to importance. The least important features may be dropped to ensure that the deadline is met. The dropped features can be implemented in a subsequent increment if they are still required.

1.7.4 Iterative model

This model (see Figure 1.3) is suited to situations in which the requirements are not clearly understood and where there is a need to begin development quickly

Figure 1.2 An incremental model

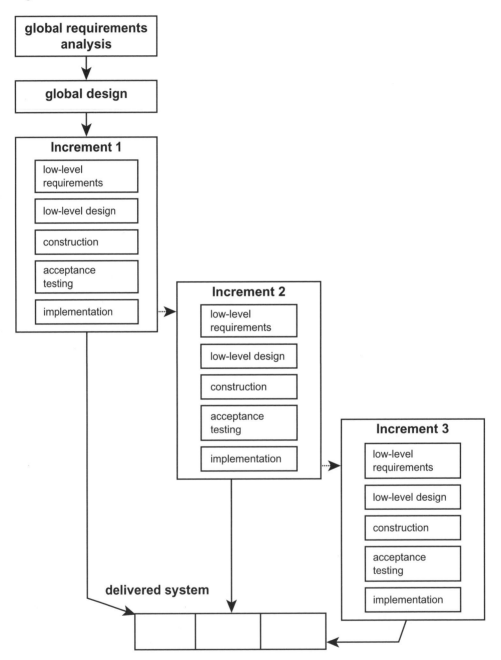

Figure 1.3 An iterative model

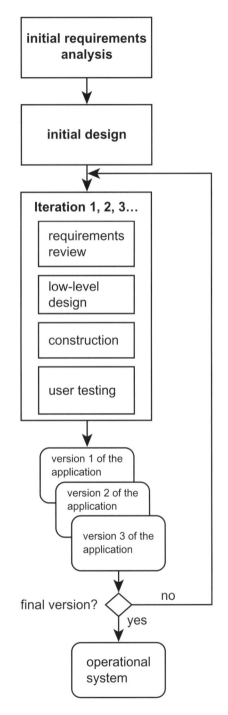

to create a version of the product which will demonstrate its look and feel. Early versions, or **prototypes**, of the system are created to help the customer identify and refine requirements and design features. The customer can make suggestions for possible changes to be incorporated into a further version of the software which is then evaluated.

A risk associated with this model is not knowing when to stop iterating. The iterative approach is potentially difficult to monitor and control.

The incremental and iterative models work well together. An application can be broken down into a number of increments, each of which can be implemented through a series of iterations.

1.8 THE PROJECT PLAN

During the project set-up phase (see Section 1.4.3), a project plan is produced that consists of several different types of documents, including activity networks and Gantt charts (see Chapter 2).

The project plan is a set of documents that co-ordinates all the various project processes, bringing together all the planning documents used to manage and control the project. It is not cast in stone and will be amended as necessary during the lifetime of the project. The plan defines the project's scope, schedule and cost, as well as the supporting processes related to risk, procurement, human resources, communication and quality.

1.8.1 Project initiation document

A good starting point is the preparation of a project initiation document (PID) or project management plan. Different project management methods give this document different names, but in essence it serves as an agreement between the sponsors and the developers of the project. It has the following key elements.

- An introduction, which describes:
    - o the project background;
    - o the document's purpose;
    - o the business justification for the project, including a brief summary of costs and benefits (see Section 1.9).
- Project goals, objectives and deliverables.
- A project organisation chart which names:
    - o the project sponsor(s);
    - o the project manager;
    - o any sub-project or stage managers;

o the lead user representative(s);

o the lead supplier representative(s), if appropriate.

The organisation chart should detail the reporting relationships of all members of the project team and identify any steering committee (or project board or project management board) established for the project. It should show the reporting relationship between the steering committee and the project team. It should also specify the levels up to which named individuals, groups (such as the steering committee) or roles can authorise:

o the commitment of resources;

o the sign-off for documentation and other project products;

o changes to goals, objectives, deliverables cost or time scale.

Chapter 8 discusses many of the issues involved in project organisation.

- A project structure section that describes how the project will be broken down into manageable portions of work which will be administered as stages.

- A list of project milestones, i.e. significant events in the project for which dates need to be clearly specified. Milestones are used to measure the progress of a project and can be the start or completion of a major project phase. Milestones are events that consume no resources in themselves, but enjoy a great deal of attention from management at a senior level. Additional resources may be used in reviewing the state of the project before or after the milestone has been reached.

- Project success and completion criteria.

- A management control section that covers:

    o the frequency, timing, recipients and format of progress reports;

    o how the plan will be produced and maintained;

    o what information will be monitored and recorded;

    o how the information will be recorded;

    o how packages of work will be signed off and reviews conducted;

    o the people responsible for recording and assessing the impact of any changes;

    o the people responsible for authorising different levels of change to goals, objectives, deliverables, cost or completion date.

Chapter 3 is devoted to monitoring and control.

- A risks and assumptions section that should identify any high-level risks to the project and propose specific actions to reduce or eliminate each risk. It is useful to include in this section a list of assumptions made in producing the report. See Chapter 7 for more on risk and its management.

- A communication plan that provides an overview of how the project will communicate with the wider business organisation, particularly with regard to changes needed in the business in order to make the implemented IT application effective.

- A report sign-off section.

The document should be signed off by staff able to represent all areas and functions committing resources to the project and those who will be affected by the project. In so doing they implicitly accept the assumptions listed. Final sign-off should be obtained from the project sponsor. The project manager should not initiate any work which has not been explicitly or implicitly authorised in the project initiation document and signed off by the project sponsor.

1.8.2 Schedule planning

Scheduling is a key project planning technique and must take place prior to the start of work. The resources required and the consequent costs will depend on the schedule decided for the work. Effective scheduling requires:

- a definition of requirements that is agreed and unambiguous;

- a careful breakdown of work;

- the creation of a coherent and internally consistent **schedule** which shows when activities will start and end and the resources that they will need;

- careful monitoring of progress against the schedule.

It is the responsibility of the project manager to make sure these requirements are fulfilled. Schedules will generally be produced at several levels of detail:

- project level, at which the project steering committee or board reviews plans and progress and takes decisions;

- phase/stage level, at which project or stage managers break down the main stages of the project into activities which are then allocated to teams;

- activity level, at which team leaders allocate work to team members so that activities allocated to the team can be completed.

As will be seen in Chapter 2, the Gantt chart is an important planning document for showing the schedule.

1.8.3 Cost planning

Once the activities have been identified, the costs they incur should be assessed. The business case for a project is based on it not costing more than a specific amount. If the costs of the project exceed the value of its benefits, the project becomes uneconomic. It is also possible for an organisation to simply run out of money for a project.

Internal IT projects are usually paid for out of a user department's budget. Where a project is being carried out for an external customer, there may be a fixed price contract. Whatever the case, it is necessary to estimate quantities and costs, set budgets and eventually control expenditure.

Effective cost planning helps:

- decision-making and control, by allowing the costs of alternative actions to be assessed;

- prompt and up-to-date collection of progress information;

- integration of expenditure projections into the project plan.

Cost planning should be done at the same levels as schedule planning; that is, at project, phase/stage and activity level (see Section 1.8.2).

Project costs may be plotted on a cumulative resource chart (see Section 3.7.2).

1.8.4 Resource planning

The project plan needs to account for various types of resources, including people, equipment and facilities. With internally resourced IT/business change projects, management must obtain resources from other parts of the company to work on a project. Functional managers responsible for various specialist departments will be key providers of resources like office space, computer equipment and specialist expertise. If you need to obtain goods or services from outside the organisation, the person responsible for contract management will be important.

A resource plan will help the project manager identify the skills needed for the project and when they are needed. For example, in many cases a project will not need a testing expert throughout its life, but only at certain times.

A useful communication tool for the project manager is the responsibility assignment matrix (RAM), a simple matrix showing individuals associated with the project on one axis and the activities for which they are responsible on the other. By plotting tasks against staff on a matrix and labelling each person's assignments, it provides a quick and easy view of who is responsible for each task.

ADVANCED TOPIC Sources of development staff

A major IT project often requires an abnormal temporary amount of technical work that the business cannot, by itself, provide. Additional development skills might be acquired by:

- Recruiting individual specialists on temporary contracts, either directly or by using agencies. Day rates for temporary staff are often greater than those for permanent staff and recruitment and training costs need to be taken into account. Depending on the type of work, temporary staff could work from their own premises, but more often they will need other facilities within the business.

- In the case above, the temporary staff are employed directly by the business. An alternative approach is to have a contract with an outside specialist company where the developers used are and remain the employees of the specialist company. The specialist company is paid for the hours that their employees work.

> • A more radical approach is to outsource one or more project activities, such as design and construction, to an outside specialist company. The outside company will have management responsibility for delivery of a subset of the project requirements, often for a fixed price. In this case, the supplier selection and deliverable acceptance processes would still consume some of the customer's resources.

1.8.5 Communication planning

Stakeholders, including the project team, have varying needs for communication about the project. These needs and the means by which they may be satisfied are recorded in a communication plan. The communication plan includes, among other things:

- the flows of communication during the project;

- how various communication tools will be used;

- what meetings will be held with what attendees, and at what times.

These issues are explored further in Chapter 8.

1.8.6 Quality planning

In order to develop a system which meets all users' functional and system performance requirements documented during analysis, a carefully considered quality plan is needed.

Quality criteria can be applied both to project deliverables and to the processes by which the deliverables are created. It is possible to check that the end products have the required qualities, and also that the processes that created them were the correct ones. Because the importance of quality often gets lost among deadline pressures and budget cuts, the project manager should emphasise it throughout the entire project life cycle, from planning to completion.

The customer must be the final judge of the product's quality, and must therefore be involved in quality evaluation. Chapter 5 further explores the role and content of the quality plan.

1.9 THE BUSINESS CASE

This is a key document for any project. As noted in Section 1.4.1, the proposal for a project may be triggered in a variety of ways. Once a proposal has been made, senior management will decide whether to go ahead with a proposal by using a combination of qualitative and quantitative criteria, as no single method gives enough information.

Qualitative criteria include **organisational fit**. Does the project fit with the organisation's strategic objectives? How does the proposed project contribute to the organisation's future capabilities and growth? The business risks associated with a particular project should also be assessed qualitatively at this point (see Chapter 7).

A more persuasive reason for going ahead with a project, however, is often **financial justification**, where two of the most common quantitative financial criteria are **net present value** and **payback period** – see Sections 1.9.1 and 1.9.2 below.

The contents of the business case document would include a description of the project, its objectives and scope, a cost benefit analysis, a risk analysis, a conceptual solution, resource requirements and success criteria. Some organisations include alternative options in the business case in order to compare the recommended solution against other approaches. The business case needs to be carefully reviewed by project sponsors.

1.9.1 Net present value

The financial business case is based on the calculation of whether the value of the benefits produced by the project exceeds the money spent on developing and installing the system and eventual costs of operation. The value of costs and income, however, is influenced by when they are incurred or received. The net present value technique looks at the benefits of having money sooner rather than later. Because money in hand can be invested to make more money, the earlier you bring money into the organisation, the more it is worth.

ADVANCED TOPIC

Calculation of Net Present Value – not needed for the BCS Foundation Certificate examination

Present value calculations translate future costs and benefits to a present day value using the formula

$$\text{Present Value} = \text{Future Value} \times (1 / (1 + i)^n)$$

where i is the interest rate (or discount rate) and n is the number of time periods (usually in years) from today. The discount rate is the rate of interest that one could expect to receive elsewhere for an investment of comparable risk.

Say we were to receive £100 in one year's time and the interest rate was 10 per cent (for ease of calculation!). The present value of that £100 would be

$$£100(1 / (1.10)^1) \text{ i.e. } £90.90.$$

In other words, if I put £90.90 into an account at 10 per cent interest then in a year I will have earned £9.09, which would give me a fraction under £100 in all.

The difference between the money generated by a delivered project less the costs in a particular year is called the **cash flow** for that year. The net present value of a project is the sum of the net present value of each annual cash flow for the delivered project. The net present value technique allows an organisation to estimate the value of money earned several years into the future and can be used to compare the expected results of investing in different projects with leaving money in a bank at a specified rate of interest.

Table 1.1 Net present value calculation (figures in £s)

Simon's company			Christopher's company		
Year	cash flow	PV	Year	cash flow	PV
0	−425	−425	0	0	0
1	400	364	1	300	273
2	450	372	2	300	248
3	500	376	3	300	225
Totals	925	687		900	746

Here is a more detailed example, which uses a discount rate of 10 per cent. Remember that for the BCS Foundation Certificate you do not need to remember the details of the calculations.

A small organisation decides to outsource its catering to outside caterers and invites companies to tender. Whichever caterer wins the contract will pay the host organisation an agreed amount of money from the revenues that it will receive from charges for meals.

The table above represents bids from the two contractors. It can be seen that Simon's company wants some money initially to help set up the operation. If you calculate the total cash flows for both proposals, then despite the initial subsidy paid to him, Simon's seems slightly better, at £925 rather than £900. When the figures are converted to reflect net present value, however, Christopher's is revealed as being much better – at £746 as opposed to £686.

COMPLEMENTARY READING

Finance for IT Decision Makers, Michael Blackstaff, BCS

1.9.2 Payback period

ADVANCED TOPIC Calculation of payback period

Net present value calculations draw attention to the fact that money received early on in the delivered system's life cycle is more valuable than that received later. It is also true that forecasts of income become progressively less accurate as we look further into the future. Interest rates could change, or the economy could go into recession and undo our calculations. Thus how quickly you begin to make money is critical. This is what payback period calculations measure. From an organisational point of view, the shorter the payback period the better. Indeed, many organisations will be looking for a payback period of one year or less.

In the Canal Dreams ebooking enhancement project, say that the expenditure to set up the online booking system is £10,000. This is allocated to year 0, which really means 'the period up to the system going live'. In the table below, the income is made up of the value of the additional bookings that the new system allows and the savings in booking staff. It can be seen the accumulated cash flow becomes positive somewhere near the middle of year 3, so the payback period is about 2.5 years.

Table 1.2 Payback period calculation (figures in £s)

Year	expenditure	income	annual cash flow	accumulated cash flow
0	10,000	0	−10,000	−10,000
1	1,000	4,000	3,000	−7,000
2	1,000	5,000	4,000	−3,000
3	1,000	7,000	6,000	3,000
4	1,000	7,000	6,000	9,000

COMPLEMENTARY READING

Finance for IT Decision Makers, Michael Blackstaff, BCS

1.10 IMPLEMENTATION STRATEGIES

At the other end of the project, there will come a time when it is necessary to convert from the old method of working to the new one brought about by the implementation of the new system. The project team, in consultation with the users, will recommend to the project board or steering committee the most suitable changeover method for the project, which may include installing IT equipment, software applications or both. The following options may be considered.

1.10.1 Direct changeover
In this case, the old system is discarded and immediately replaced by the new one. It can be considered a risky approach, but is relatively inexpensive if thorough testing has been done. The more complex and important the new system, the riskier this approach will be, especially if there is no possibility of falling back on the old system in the case of failure.

1.10.2 Parallel-running
This involves running the old and new systems together for a period of time using the same inputs and comparing the related outputs – so it serves as a continuation of the testing process. It is a safe, low-risk approach but can be expensive,

particularly in terms of duplicated labour costs. An important management decision is how long to run the two systems in parallel.

1.10.3 Phased take-on

The phased approach breaks the system into components that will be introduced in sequence. It helps minimise risks but can delay the implementation of the entire integrated system. However it does present the opportunity to allow users to learn one system component at a time. It also fits neatly with incremental delivery.

1.10.4 Pilot changeover

Like the phased approach, this is a risk-reducing approach. With pilot changeover the entire new system is introduced to just one business unit or location. It can only be used if the business unit or location can use the entire system independently. Problems can be addressed and fixed before the system is introduced company-wide, but company-wide deployment of the entire system is consequently delayed.

ACTIVITY 1.4

What would be the best implementation strategy for the Canal Dreams ebooking enhancement project?

1.11 POST-IMPLEMENTATION REVIEW

The **post-implementation review (PIR)**, or **project evaluation review**, is usually scheduled to take place some 6 to 12 months after the sign-off of the project. Its objective is to review the implemented system in terms of its contribution to business objectives, its usability, operating costs and reliability. It considers the following:

- whether the business and system requirements have been met;
- cost and benefit performance;
- operational performance;
- controls, auditability, security and contingency;
- ease of use.

The output from this process is a **post-implementation review report**. The review should be led by someone who is independent of the project and should solicit feedback from users, operations and the support team. The review should address the operational system and not the development project. The additional effort required of them might lead to users not engaging with this review. If this is the case, it is essential to explain to the users the benefits of the process – for example, that changes that could improve the system may be identified as a result.

The PIR report needs to be distinguished from a **lessons learnt report**. When the project has been delivered by the project team, the project manager should write a report describing the major challenges experienced during the project and how they were managed. The purpose of this report is to identify lessons that would improve the execution of future projects.

SAMPLE QUESTIONS

1. Which of the following is NOT a characteristic of a project?
(a) Ongoing nature
(b) Uniqueness
(c) Clear objectives
(d) Integration of interrelated tasks and resources

2. Which of the following is NOT managed by the project manager?
(a) Time, cost and scope
(b) The project team
(c) The project sponsor
(d) Expectations of the stakeholders

3. Which of the following tools indicates who is responsible for what?
(a) A responsibility assignment matrix
(b) A resource levelling chart
(c) An activity network chart
(d) A resource histogram

4. With which one of the following does the calculation of the payback period provide the organisation?
(a) An assessment of how quickly an implemented system will produce a profit
(b) How much future income from an implemented project will be worth in present day terms
(c) The discount rate that would give a payback with a net present value of zero
(d) The overall income that the implemented system should produce, less any initial investment

ANSWERS TO SAMPLE QUESTIONS

1. (a) 2. (c) 3. (a) 4. (a)

POINTERS FOR ACTIVITIES

ACTIVITY 1.1

Among the activities that may be considered are the following.

(i)   Survey existing office requirements (for example, how many desks and chairs there are) and IT infrastructure, including servers and printers, used by the office.

(ii)   Survey new office space.

(iii) Plan new office layout – who and what will go where?

(iv)   Schedule the sequence of moves – it might be that not all staff should be moved at once as this will allow for some continuity of service.

(v)   Select a removal company.

(vi)   Organise the setting up of the infrastructure.

(vii) Pack up the old office.

(viii) Transfer to the new office, including supervision of placement of furniture etc.

(ix)   Unpack in the new office.

(x)   Connect telephones, etc.

As project manager you will probably want to be consulted when decisions have to be made, possibly at points (iii), (iv) and (v) as money is involved here. Useful checkpoints would be just before (vii) to check everything is in order to go ahead and after (x) to check out any outstanding problems.

ACTIVITY 1.2

There will be some interviews with fairly high-level managers, including the project sponsor (who may be the managing director, who also owns the company), to clarify the business requirements of the proposed system. Given the nature of this project. marketing specialists will need to be consulted. It is essential that this consultation take place when the business case is being prepared.

Staff who currently carry out the clerical booking operations would need to be interviewed to see how the existing system works, what data has to be held and what problems there are with the current system. Operations staff at boatyards also have to be consulted as users of the new system who need information about the bookings each week so that boats can be allocated and prepared and holiday-makers checked in. Staff in other departments need to be approached to document the interfaces between the booking operation and other parts of the overall Canal Dreams business, for example the central finance function and the maintenance staff who may need to schedule boat maintenance. In many cases, existing functionality is unaltered, but performance requirements may change. For example, the move towards online bookings may increase the number of online debit and credit payments.

Note that users include IT operational staff who would, for example, have requirements about system security as well as concerns about the move to a 24/7 service.

We usually say that end-users should be consulted, but of course, in internet transactions with the public, it is not possible beforehand to identify who these will be. In this case market research with recent telesales customers, for example, could provide insights into the needs of online customers.

ACTIVITY 1.3

Although there will be the existing functionality used by telesales, functions where potential customers have to 'volunteer' to use the system need more careful design than those which employees are compelled to use. It is assumed that these external transactions will be the subject of redesign to ensure their ease of use:

- Browse locations and boats,

- Check availability,

- Record boat booking,

- Cancel boat booking,

- Change allowable details, for example amend existing address, and 'crew list'.

In order to make an internet booking, the customer must have internet access. Other communications that were previously sent by post could now be sent by email. These would include booking confirmation, final payment reminders and proof of booking and other details needed when starting the cruise.

ACTIVITY 1.4

The canal holidays business is seasonal, with very busy times of the year and times in the off-season when the business is dormant. This would seem to favour a direct changeover during a quiet period. This would allow a gradual build-up of traffic, making it easier to deal with any initial teething troubles. Note, however, that direct changeover is often avoided because of the risk that if the new system is faulty there is no alternative system to fall back on.

2 PROJECT PLANNING

LEARNING OUTCOMES

When you have completed this chapter you should be able to demonstrate an understanding of the following:

- *project deliverables and intermediate products;*
- *work and product breakdowns;*
- *product definitions (including the identification of 'derived from' and 'component of' relationships between products);*
- *relationship between products and activities in a project;*
- *checkpoints and milestones;*
- *elapsed time and effort required for activities;*
- *activity networks (using the 'activity on node' notation);*
- *calculation of earliest and latest start and end dates of activities and the resulting float;*
- *identification and significance of critical paths;*
- *resource allocation, smoothing and levelling, including the use of resource histograms;*
- *work schedules and Gantt charts.*

2.1 INTRODUCTION

In this chapter we describe the main steps in producing an initial plan for a project. You will recall from Chapter 1 that before the detailed planning of a project starts, the **business case** for the project should have been set out. This shows how the value of the proposed IT application's benefits are expected to outweigh the costs of developing and managing it. The overall **objectives** of the project, which define the successful outcomes of the project, have also been identified and have been agreed by the main participants in the project.

2.2 APPROACHES TO PLANNING

There are two approaches to identifying the components of a project: **product-based** and **work-** or **activity-based**.

2.2.1 Product-based planning

With the product-based approach, detailed planning usually begins with identifying the **project deliverables**: that is, the products that will be created by the project and delivered to the client. A product must be in some way tangible, but it can be any of a wide range of things. It could be a software component, a document, a piece of equipment or even a person (for example, a trained user). It could be a new version of some existing product, as with a modified version of a software component.

In the case of the Canal Dreams ebooking enhancement project, the deliverables include:

- software functionality which enables members of the public to book canal holidays via the internet;

- enhanced network systems that can cope with the expected additional traffic to the Canal Dreams website;

- a new network support centre to support 24-hour/seven days a week activity.

Once the deliverables have been defined, **intermediate products** can be identified. These are products which are created during the course of the project, but which may not actually be delivered to the client at the end of the project. In the case of the Canal Dreams ebooking enhancement project, intermediate products include (among other things):

- business process models;

- interface design documents, such as a corporate website style guide, site maps, 'wire-frame' designs,

- an enhanced IT infrastructure architecture plan;

- software specifications;

- acceptance test plan;

- progress reports.

Some products, such as the progress reports in the list above, will relate to the management or quality control of the project.

The deliverable and intermediate products can be written simply as a list of products, but sometimes they are shown in a **product breakdown structure** diagram.

Some of the stakeholders in the project may find there are some products with which they are unfamiliar. Some users, for example, may be unsure of what is meant by an 'acceptance test plan'. To remedy this, planning should include drawing up **product definitions**. For each product, the following should be documented:

- The **identity** of the product, for example 'acceptance test plan'.

- A **description** of the product, for example 'a plan of the test cases and the results that the users expect the application to produce'.

- The product or products that have to exist before this one can be created: that is, those it is **derived from**. For example, the acceptance test plan is stated to be derived from the requirements specification, which describes the main transactions of the application.

- The **components** that make up the product – in the case of an acceptance test plan, the main sections in the document.

- The **format** of the product – for example, that it is a word-processed document or a spreadsheet or a piece of software.

- The **quality criteria** that explain how the product will be judged as satisfactory – for example, the acceptance test plan being reviewed against the requirements specification.

2.2.2 Work and product breakdown structures

An alternative method of planning is the work- or activity-based approach, which identifies the required work activities or tasks in a work breakdown structure (WBS). In this case, the intermediate products related to setting up the Canal Dreams ebooking project listed above would be replaced by activities such as:

- analysing/redesigning business processes;

- designing web interfaces;

- redesigning network architecture;

- specifying software;

- planning acceptance tests;

- reporting progress.

As nearly all activities will generate a product – or else why do them? – and all products will need to have some activities that give birth to them, there may not really be much difference between the two approaches in practice.

Figure 2.1 A product breakdown structure diagram

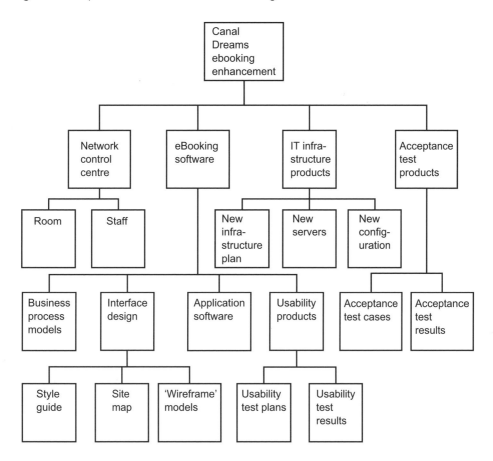

ACTIVITY 2.1

Which products are created by each of the following?
a) testing
b) training
c) network installation
d) a project progress meeting

2.3 PRODUCT FLOW DIAGRAM

If you have adopted a product-driven approach, it is possible to draw up a product flow diagram (PFD) showing the order in which the products have to be created. This should be relatively easy to draft if you have already produced product definitions that specify from which other products each product is derived. Figure 2.2 gives an example fragment of a product flow diagram.

Figure 2.2 A product flow diagram

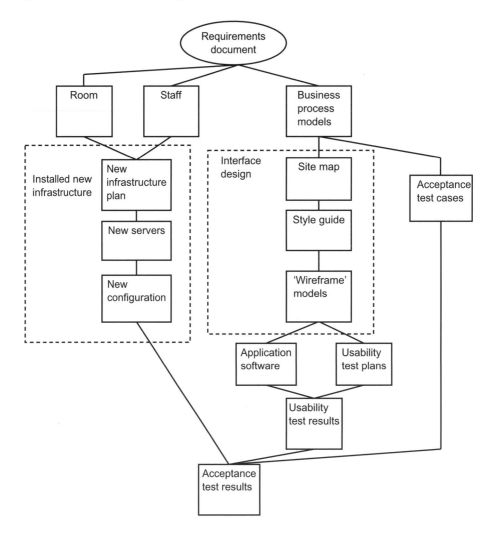

Note the oval with 'requirements document' in it. This refers to a product that already exists and that will be used to generate one or more of the products in the PFD. There is no one correct PFD – its structure depends on the decisions made. For instance, in Figure 2.2 the assumption is that additional network specialists will be recruited who can work on enhancing the current IT infrastructure, and then support the new operational system. An alternative would be to bring in network specialists on short-term contracts to do the enhancement work. The support of the new infrastructure could be carried out by different people – perhaps someone already working for Canal Dreams in a now-redundant role could be retrained.

The flow of a PFD is normally from top to bottom and then left to right. Looping back is not allowed – not because this cannot happen in real life, but because it can almost always be possible **technically** to go back and rework a product previously thought to be completed. In this case all the products depending on the reworked one might also need some reworking.

In two places in Figure 2.2 we have put boxes of broken lines around a sequence of products. This is not part of the official PFD notation: we wish to show that a group of components – in one case, site plan, style guide and 'wire-frame' designs – will be treated as one large product – interface design.

2.4 ACTIVITY PLANNING

Whether a product flow diagram has been drawn up or whether the planner has simply drawn up a list of activities, the next step is to draw up an **activity network**. This shows the activities needed and the order in which they are to be carried out.

2.4.1 Activity network diagram
There are two sets of conventions for drawing up activity networks: 'activity on node' and 'activity on arrow'. Figure 2.3 shows an example of an activity on arrow diagram.

Figure 2.3 Activity on arrow network

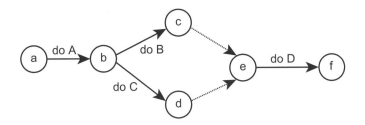

As the name implies, the arrows in an activity on arrow diagram represent activities, while the circles that link the arrows (that is, the **nodes**) represent the ends of some activities and the starts of others. The arrows with broken lines indicate 'dummy activities' which simply show a dependency between two of the event nodes – for example c, the end of 'do B', and e, the start of 'do D'.

In these notes we use a different set of conventions, **activity on node**, which is used by most modern project planning tools, including Microsoft Project.

Figure 2.4 shows the same activities as Figure 2.3, but using activity on node notation. Here the boxes (which are the 'nodes' in this case) represent activities while the lines between the boxes show where the start of one activity depends on the completion of some other activity. Note that at this stage the constraints may

Figure 2.4 Activity on node network

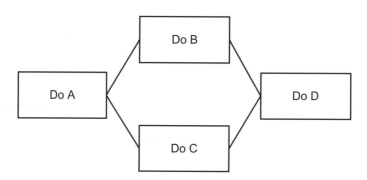

be technical or external. A technical constraint normally means that a product has to be created by one activity so that another can use it. An example of an external constraint is a system that can only be tested out of office hours, so it has been agreed contractually that testing will take place at weekends only.

What are not taken into account at this point are **resource constraints** – for example, that a person will not be able to start on one task because he or she will not have finished another. These considerations are deferred because a decision might be taken, when staff are being allocated, to employ more staff.

ACTIVITY 2.2

In this activity network, match the activities with the boxes so that the activity network is compatible with the product flow diagram in Figure 2.2.

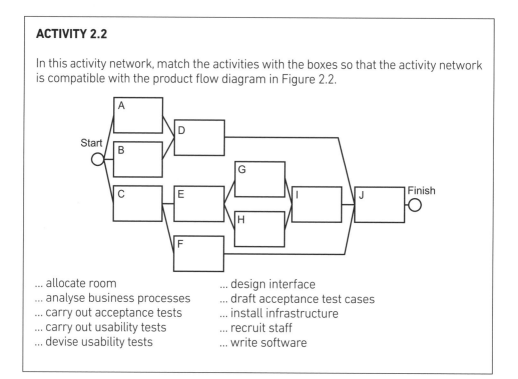

... allocate room
... analyse business processes
... carry out acceptance tests
... carry out usability tests
... devise usability tests

... design interface
... draft acceptance test cases
... install infrastructure
... recruit staff
... write software

In Activity 2.2 we added a 'start' and 'finish'. These are important points of time (or **'events'**) in the life of the project, but they will not actually take up any time. If, for example, the finish of the project was marked by a celebration that took up several hours, then the 'event' would become an activity in its own right. We call these important events **milestones**. Milestones can also be located in the middle of a project, for example at the end of one important phase and the start of the next. Always remember, though, that milestones do not take up time. Sometimes an important point in a project may be marked by a meeting to check that everything planned has been completed successfully before the next part of the project starts. This checkpoint would be an activity in its own right.

2.4.2 Estimating elapsed time

Having identified the activities and the order in which they have to be worked on, we now need to estimate how long we think each activity is likely to take. Note that we are concerned here about **elapsed times**. This is the time from the start of an activity to the finish. This is not the same as the **effort** spent on an activity. Effort could be more than elapsed time – for example, where we have three people working on a job for two days, the elapsed time would be two days but the effort would be six staff days. Effort could also be less than elapsed time – for example, where someone works only afternoons. Estimates of effort become important when cost planning is being done – recall Section 1.8.3.

Let us assume that we can allocate estimated durations to the activities in the activity network of Figure 2.4 (see Figure 2.5).

We want to calculate the earliest day upon which each activity can start. Rather than worry about taking account of weekends and public holidays at this point, we simply allocate each day a sequence number, starting with day 0. (Technically, day 0 means 'the end of day 0', which means the start of day 1, as explained below.)

The **earliest start date** for 'Do A' is day 0 by definition, because it is the first activity in the network. The earliest time at which the activity can finish is day 0 plus the duration of the activity: that is, day 4.

$$\textbf{earliest finish date} = (\textbf{earliest start date} + \textbf{activity duration})$$

Figure 2.5 A network activity fragment with activity durations

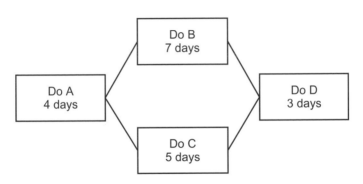

The earliest start dates for the two activities 'Do B' and Do C' are governed by the earliest finish date of the preceding activity, 'Do A'. In fact, we can say that in this case the earliest start dates for 'Do B' and 'Do C' are the same as the earliest finish date for 'Do A'.

You may wonder why it is not the **following** day. Well, the convention is that when we say 'Do A' finishes on day 4, we mean at the **end** of day 4. When we say 'Do B' and 'Do C' start on day 4 we really mean at the **end** of day 4, which of course really means the start of day 5. It is best just to accept this as the convention. It saves problems arising where activities do not take whole numbers of days, for example 5.5 days.

We can now work out the earliest finish days of 'Do B' and 'Do C' as day 11 and day 9, respectively. What about the earliest start date for 'Do D'? We have two preceding earliest finish dates, so we take the one which is later: that is, day 11.

earliest start date = the latest of the earliest finish dates of the preceding activities upon which the current activity is dependent

We end up with the day numbers shown in Figure 2.6, where ES means the earliest start date and EF means the earliest finish date.

Figure 2.6 Earliest start and finish days *(ES = 'earliest start'; EF = 'earliest finish')*

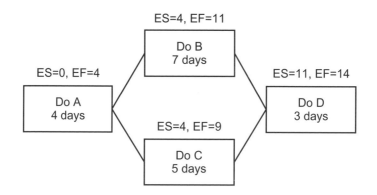

It is possible for some activities to start or finish late without the project as a whole being delayed. To see where this is the case, the **latest finish** and **latest start** dates for each activity are calculated. We will assume that we want the project as a whole to take the shortest time possible: that is, to finish on day 14. Day 14 becomes the latest finish date for the activity 'Do D'. The latest start day for this activity is calculated by subtracting the duration from the latest finish: that is, $14 - 3 = \text{Day } 11$.

latest start date = latest finish date of current activity − duration

We now work backwards. The latest start day for the activity 'Do D' becomes the latest finish day for 'Do B' and 'Do C'. By subtracting the durations for these activities from their latest finish days we get their latest start days: that is, $11 - 7 = $ Day 4 for 'Do B' and $11 - 5 = $ Day 6 for 'Do C'.

In the case of 'Do A' we have to decide whether to base the latest finish on the latest start date of 'Do B' or 'Do C'. The earlier of the two is taken as the latest finish time for 'Do A' – that is, day 4 – which comes from 'Do B'.

latest finish date = the earliest of the latest start dates of the activities that are dependent on the current activity

We now have the situation shown in Figure 2.7, where LS means latest start and LF means latest finish.

Figure 2.7 Latest start and finish dates

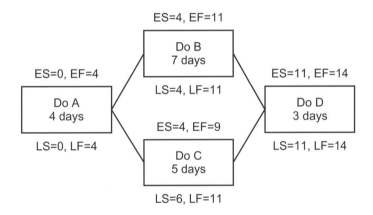

It can be seen from Figure 2.7 that for all the activities except 'Do C', the earliest and latest **start** days are the same, as are the earliest and latest **finish** days. This means that if these activities are late, the project as a whole will be delayed. In the case of 'Do C', if you look at the day numbers you can see there is a difference of two days between the earliest and latest day numbers. This means that 'Do C' could be one or two days late and the duration of the project as a whole would not be affected.

This leeway is called the **float** and can be defined as:

float = latest finish date – earliest start date – duration

A quick way of calculating this is by subtracting the earliest start from the latest start (or the earliest finish from the latest finish).

In Figure 2.7, 'Do A', 'Do B' and 'Do D' all have zero float. They form a small chain of three activities from the beginning to the end of the activity network. This chain is the **critical path**. If any activity on this path is delayed then the whole project will be delayed.

The details for each activity can be displayed more clearly if the boxes on the activity diagram are divided up as shown in Figure 2.8.

Figure 2.8 Layout of an activity box

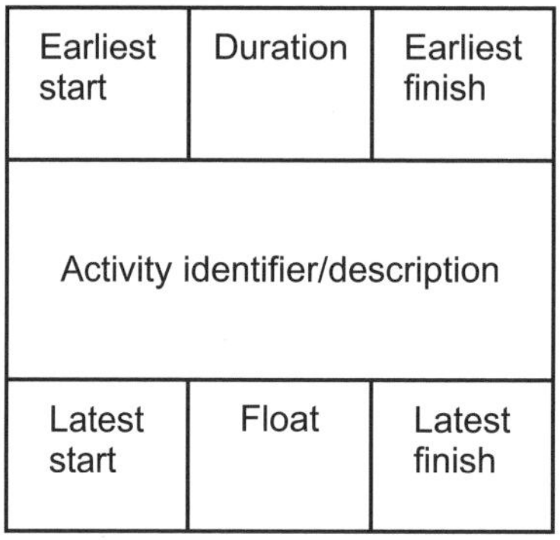

Earliest start	Duration	Earliest finish
Activity identifier/description		
Latest start	Float	Latest finish

Thus for the activity 'Do C', the activity box in the activity network could be drawn up as in Figure 2.9.

Figure 2.9 Activity box for 'Do C'

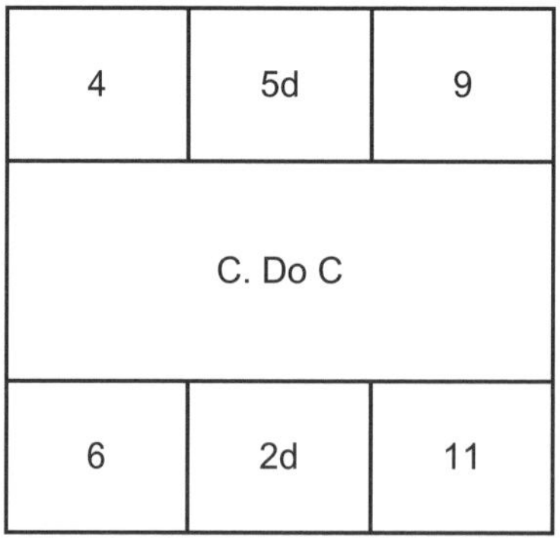

4	5d	9
C. Do C		
6	2d	11

The **activity span** is the total period during which the activity has to take place, and is defined as:

activity span = latest finish day − earliest start day

In this case it is 11 − 4: that is, 7 days. Where an activity has float, there is a 'window of opportunity', reflected by the activity span (see Figure 2.10), within which the activity can start and be completed.

Figure 2.10 The activity span

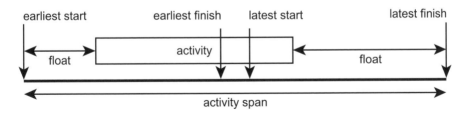

The activity has to take place within the activity span. Because there is some float, there is some freedom about when it can take place within that period. However, the start must be in the period between the earliest and the latest start. If it is not, it will not be completed within the activity span – unless its duration can be shortened in some way.

ACTIVITY 2.3

Calculate the earliest and latest start and finish weeks and floats for each of the activities in the activity network below. Use the results to identify the critical path.

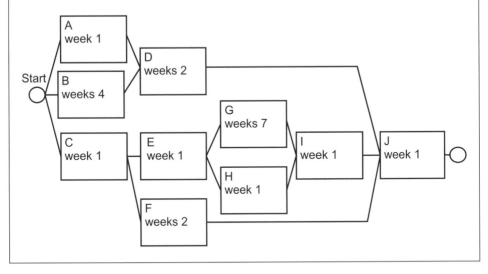

In Section 1.7.4 the iterative model was introduced, where one or more activities could be repeated, with each loop creating a new version of the products of the process. Activity networks assume that all activities are executed just once. Clearly, some – such as usability tests – can be carried out more than once on revised versions of the software. We deal with this by concealing the iteration within an activity. Usability tests may be carried out a number of times on revised versions of the software, but all the iterations are together expected to take no longer than one week.

2.5 RESOURCE ALLOCATION

So far we have taken no account of the availability of the resources needed to carry out any task. It is assumed that they will be available when they are needed. The resources that now need to be considered include raw materials, staffing and equipment. Usually with IT projects, the main concern is staffing; however, sometimes equipment can also cause problems. For example, when conducting acceptance testing for a modified IT application, there is often a need for some testing of the whole operational system. This may need to be done on a public holiday or at the weekend, when no normal operational use is being made of the system. Here, we focus on staff resourcing.

For each activity, the **resource types** needed are identified. A resource type is a group of people of which any member could carry out a particular task. For example, if a software component needs to be written in Java, identifying Ali as the needed resource would be too precise; Jane may be equally proficient in Java. Identifying the resource simply as a software developer, on the other hand, may be too vague: Alfred is a software developer but, as a COBOL programmer, he has no knowledge of Java. Identifying the required resource as a Java programmer may be just about right.

ACTIVITY 2.4

Match the following resource types and activities. More than one resource type might be needed for an activity.

Activities	Resource types
allocate room	business analyst
analyse business processes	human resources manager
carry out acceptance tests	interface designer
carry out usability tests	IT infrastructure support
devise usability tests	premises manager
design interface	software developer
draft acceptance test cases	users
install infrastructure	IT manager
recruit staff	
write software	
….	

In order to illustrate the process of resource levelling and smoothing, we break down one of the activities in our Canal Dreams ebooking enhancement project, 'write software', into more detailed tasks (see Figure 2.11).

Having allocated resource types to activities, we now go through the activity network and, for each unit of calendar time (in this case each week), note the resources needed if the activity is to start at its earliest start date. In the top part of Table 2.1 we have, for example, identified the different resource types and allocated particular activities to them by putting the alphabetic characters we used in Figure 2.11

Figure 2.11 Canal Dreams project: write software activity
(SDes = System designer, Prog = Programmer, w = week)

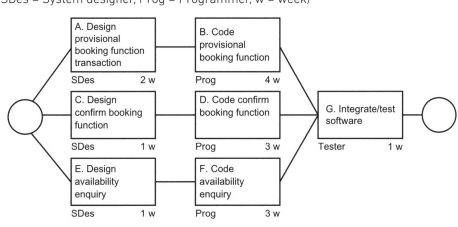

Table 2.1 Numbers of each resource type needed in each week

Week >	1	2	3	4	5	6	7
Resource	Activity they have been allocated to in each week						
System designers							
SDes 1	A	A					
SDes 2	C						
SDes 3	E						
Programmers							
Prog 1			B	B	B	B	
Prog 2		D	D	D			
Prog 2		F	F	F			
Testers							
Test 1							G
Number of each resource type needed in each week							
System designers	3	1	0	0	0	0	0
Programmers	0	2	3	3	1	1	0
Testers	0	0	0	0	0	0	1

to identify each activity e.g. 'A' for 'design provisional booking function' into the relevant week cells. Where the same type of resource is needed for different activities in the same week, we identify different instances, for example SDes 1, SDes 2 and so on and allocate them to the different parallel activities where they are needed. When we have finished this allocation we can count the number of each type of staff needed in each week. We can also depict this information as a **resource histogram** (see Figure 2.12).

Figure 2.12 A resource histogram for each resource type

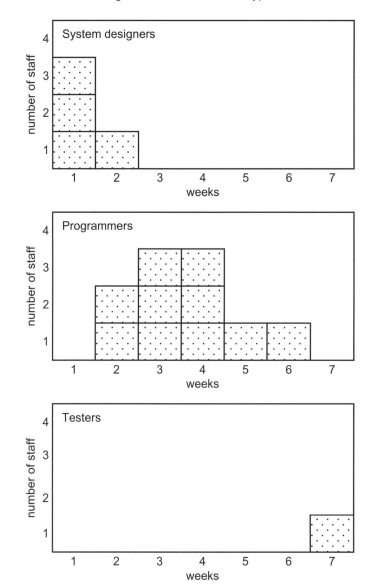

In some cases, we may find that the project plan needs more staff of a certain type than we have available. Another concern is that where we are acquiring staff from outside, we want to make sure that, as far as possible, we are able to provide a steady flow of work. Where temporary staff need to be employed on expensive contracts, we try to avoid having short periods of intense activity alternating with periods when nothing much is happening and staff are underemployed. As well as the additional costs, temporary staff's lack of familiarity with the project may mean that they are less productive at first.

Delaying certain activities may allow peaks and troughs to be 'smoothed', which will lead to more economical staff costs and more productive work distribution. For example, in Figure 2.12, three system designers are needed in week 1 and only one in week 2. If we have only two system designers on the staff, then we could employ a third system designer, possibly on an expensive temporary contract, for the first week. If we calculate the float for each activity in the activity network (Figure 2.11), we can see that two of the activities needing system designers have two weeks' float. If we delay starting one of these activities until week 2, this part of the project as a whole will not be delayed, but we will need only two system designers.

When there are not enough staff of a particular resource type to carry out the activities due to take place at a certain time, there is a **resource clash**. Even if there are no resource clashes, a planner should try to arrange activities so that the usage of a resource type is as stable as possible.

ACTIVITY 2.5

Redraw Table 2.1 to take account of a delay of one week to the activity 'design availability enquiry'.

Where there is a resource clash or the demand for a resource is very uneven, the following options may be considered:

- Use the float of an activity to delay the start of some work until the required staff member is available (as in the example above).

- Delay the start of an activity even though the float has been used up. This will delay the overall completion date of the project, but that may be preferable to the extra cost of employing more staff.

- Buy in additional staff to cover the staff deficiency. This will normally increase the cost of the project.

- Split an activity into sub-activities. For example, it may be possible to split the provisional booking function into two component sub-functions, each requiring a week of design and two weeks of programming. This could allow the demand for systems design to be spread more evenly.

ACTIVITY 2.6

Redraw Table 2.1 to reflect the demand for the different types of resource if the provisional booking function is split into two equal-sized software components, each needing two weeks of coding, and a management decision is made to employ only one systems designer.

We are now in a position to put our plan into a form that will be easily understood by all those who are going to be carrying it out. The most common format used is a Gantt chart (see Figure 2.13). The activities are listed down the left-hand side and the calendar units (in this case, weeks) are shown along the top. In the body of the diagram there are block symbols for each activity, showing when the activity will be carried out. In the diagram, the **free float** related to each activity is indicated by the lighter blocks that extend the base period for an activity. Free float is the amount of time that an activity can be late without any other activity being late. For Activity A (Allocate room), this is 3 days. If Activity A is later than 3 days, the start of Activity D (Install infrastructure) will be delayed and Activity D's free

---

Figure 2.13 Gantt chart

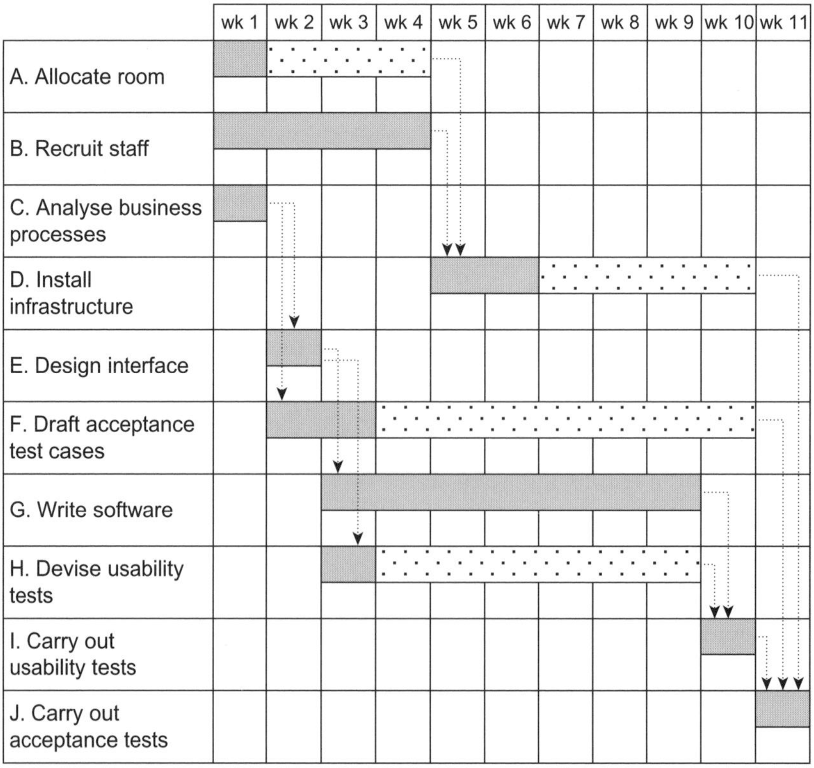

float will be reduced. However Activity A can be up to 7 days late and, as long as Activity D does not take longer than planned, the project as a whole will not be delayed. The 7 days is known as Activity A's **total float** and is what appears on the activity network.

For a smaller project, an alternative layout is to list each member of staff down the left and show what they are doing in each time period in the body of the diagram. This is not dissimilar to the holiday planning charts on the walls of many offices showing when staff will be away. A disadvantage of this format is that activities involving more than one team member have to be duplicated.

If an activity has to be delayed until a member of staff becomes available upon the completion of some other activity, this should **not** be indicated by a dependency link between the two activities on the activity network. Rather, the start date of the waiting activity should simply be amended on the Gantt chart. If other staff were to be released earlier than foreseen they could be used to expedite the waiting activity. If the waiting task had a dependency link to another task, it would imply a technical reason for waiting until the other was complete and would mask the opportunity to use other staff.

ACTIVITY 2.7

What does GANTT stand for in the name 'Gantt chart'?

2.6 USING SOFTWARE TOOLS FOR PLANNING

In this chapter it has been assumed that all the planning, and calculation of the consequences of particular planning decisions, will be done by hand with no computer assistance. It will be a relief for most people to know that there are software packages that will carry out most of the calculations for you. Examples of these are Microsoft Project and Oracle Primavera.

In most cases, for each activity you will input the activity name, the duration of the activity, the activities upon which it is dependent and the resources that it will use. Given this information, the software then produces activity networks, resource histograms, Gantt charts and other useful reports. Some packages suggest ways of resolving resource clashes, but users need to check that the results are what they really want. Where activity networks and Gantt charts are produced, the planner often has to tweak them to make them easy for others to understand. For example, a Gantt chart often spreads over several pages, many of which are blank. The advantage of using a software tool is that it is easy to make changes to the plan and see what the consequences of the changes will be.

SAMPLE QUESTIONS

Questions 1, 2 and 3 are about the diagram below. The number of days in each box show the duration of the activity.

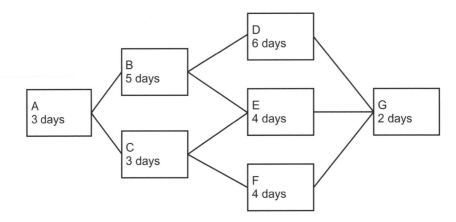

1. Which of the following is the critical path?
(a) A, B, D, G
(b) A, B, E, G
(c) A, C, E, G
(d) A, C, F, G

2. What is the float for activity F?
(a) 0 days
(b) 2 days
(c) 4 days
(d) 6 days

3. Activities B and C have to be completed by the same person. What is the delay in the end time of the project?
(a) 3 days
(b) 5 days
(c) 4 days
(d) 1 day

4. Which of the following does not take account of the dependencies between activities?
(a) Gantt chart
(b) activity network
(c) work breakdown structure
(d) resource histogram

ANSWERS TO SAMPLE QUESTIONS

1. (a) 2. (c) 3. (d) 4. (c)

POINTERS FOR ACTIVITIES

ACTIVITY 2.1

Among the products that may be created for each activity are:

(a) test results, error reports
(b) trained users
(c) a new network
(d) meeting minutes, to-do lists, updated plans

ACTIVITY 2.2

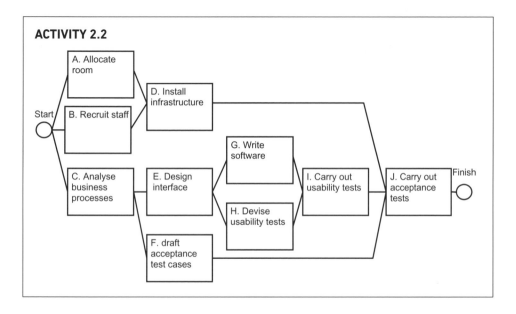

ACTIVITY 2.3

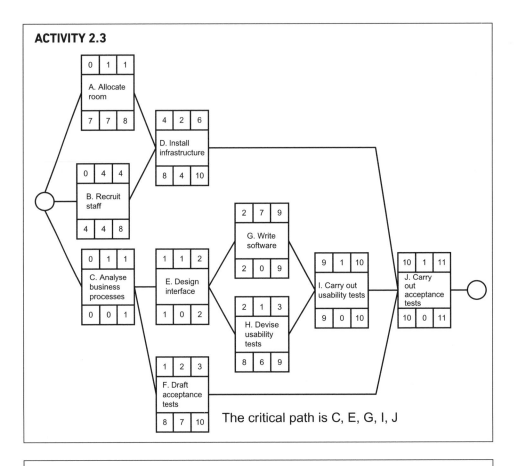

The critical path is C, E, G, I, J

ACTIVITY 2.4 ONE WAY OF ALLOCATING RESOURCES WOULD BE:

Activity	Resource type
Allocate room	Premises manager
Analyse business processes	Business analyst, users
Carry out acceptance tests	Users, business analyst
Carry out usability tests	Interface designer, users
Devise usability tests	Interface designer
Design interface	Interface designer, users
Devise acceptance test cases	Business analyst, users
Install infrastructure	IT infrastructure support
Recruit staff	Human resources manager
Write/test software	Software developers

ACTIVITY 2.5

Week >	1	2	3	4	5	6	7
Resource	Activity they have been allocated to in each week						
System designers							
SDes 1	A	A					
SDes 2	C	E					
Programmers							
Prog 1			B	B	B	B	
Prog 2		D	D	D			
Prog 2			F	F	F		
Testers							G
Test 1							
Number of each resource type needed in each week							
System designers	2	2	0	0	0	0	0
Programmers	0	1	3	3	2	1	0
Testers	0	0	0	0	0	0	1

ACTIVITY 2.6

Week >	1	2	3	4	5	6	7
Resource	Activity they have been allocated to in each week						
System designers							
SDes 1	C	E	A	A			
Programmers							
Prog 1		D	D	D	B(i)	B(i)	
Prog 2			F	F	F		
Prog 2				B(ii)	B(ii)		
Testers							
Test 1							G
Number of each resource type needed in each week							
System designers	1	1	1	1	0	0	0
Programmers	0	1	2	2	3	2	0
Testers	0	0	0	0	0	0	1

ACTIVITY 2.7

GANTT does not stand for anything. Gantt charts are named after their inventor, Henry Gantt. Gantt should, therefore, not be written in capitals!

3 MONITORING AND CONTROL

LEARNING OUTCOMES

When you have completed this chapter you should be able to demonstrate an understanding of the following:

- *the project control life cycle, including planning, monitoring achievement, identifying variances and taking corrective action;*

- *the nature of and purposes for which project information is gathered;*

- *how to collect and present progress information;*

- *the reporting cycle;*

- *how to take corrective action.*

3.1 INTRODUCTION

Chapter 1 described the typical stages of a project that implements an information system. The importance of controlling the project to ensure that it conforms to the plan was stressed. In Chapter 2, the way in which the plan for a particular project is created was explained. This chapter explores the means by which a project is monitored and controlled so that it broadly fulfils its plan. The mechanism for this is the project control life cycle.

3.2 THE PROJECT CONTROL LIFE CYCLE

The project control life cycle involves the following sequence of steps:

(a) Producing a plan for the project to follow;

(b) Monitoring progress against the plan;

(c) Comparing actual progress with the planned progress;

(d) Identifying variations from the plan;

(e) Applying corrective action if necessary.

Steps (b) to (e) are repeated to continue the **control cycle**.

Imagine a ship's voyage across the Channel from Dover to Calais as your project. The plan would probably involve following a certain route, aiming to arrive in Calais at a certain time. As the voyage progressed, the navigator would check the ship's progress against the planned course. If there was a difference, he or she could then decide that a change of speed or an alteration of course was necessary – this would be corrective action. The process would, of course, continue until the ship arrived at its destination. Without this control cycle, the ship could continue on a fixed course and speed and would be very unlikely to arrive at the planned destination or at the expected arrival time.

Monitoring progress is more difficult in an IT project than in the ship example. The first question, which we tackle in the next section, is how to identify things that should be monitored. We usually know what the final objective of the project is, but how do we know how well we are progressing towards that objective?

3.3 MONITORING PROGRESS

3.3.1 What should we monitor?

The most obvious thing to monitor is the **progress** in creating **deliverables** and other, **intermediate**, project products, and in meeting **milestones** or **deadlines**. Difficulties however arise when you want to monitor progress and things are only partially complete. The simple answer is to break the products and deliverables into smaller components that can be assessed as complete at shorter and more frequent intervals of time – for example, software can be broken down into smaller, relatively self-contained modules.

Where this is difficult, an alternative is to assess the percentage completion of an activity or deliverable. This can be problematic. If someone is building a wall, it is easy to see when it is half finished, although even then the finishing touches may take a little extra time. In software development it is quite common to take a long time to iron out the final snags in an 'almost finished' product. Software is also less obviously visible than a wall.

In Chapter 6, we describe **size** or **effort drivers**. These allow us to measure the size of the job to be done. In the case of building the wall, the number of bricks would be an obvious size driver: the bigger the wall, the more bricks it will need. The size/effort driver can be used to monitor progress. For example, if we know that the bricklayer will need to lay 200 bricks to build the wall but only 50 have been laid so far, then we can assume that the job is about 25 per cent complete.

The **use of resources** also needs to be monitored, which in IT projects means mostly 'human resources' or staff time. Also, **financial expenditure** should be carefully monitored. In the scenario in Activity 3.1, allowing the installer to stay in an hotel between installations in the same region may save on travelling time (and fuel costs) and speed up the installation rate, but it would need to be balanced against the additional cost of accommodation. Surprisingly, however, financial expenditure on human resources is not always strictly monitored in IT projects if

---

ACTIVITY 3.1

There are 20 boatyards owned by Canal Dreams. As part of the enhanced ebooking system, online customers will be emailed an e-ticket, containing a barcode, which they will be expected to present at the boatyard at the start of their holiday with evidence of their identity. The e-ticket requires new IT equipment to be installed at each boatyard. It has been estimated that the installer will, on average, need a day to travel to a boatyard, install the new equipment and show local staff how it is used. Twenty days (or four working weeks) have been allocated for the installation of all the equipment.

However, at the end of the first week only three boatyards have in fact been visited.

(a) How long is it likely that the installation programme will now take?

(b) What difference to the figure you have produced in (a) may be made by the following circumstances?

   (i) The installer started two days late because some items of equipment had not been delivered;

   (ii) The installer started with the boatyards furthest afield and needed an extra day to travel to the area and back.

---

the project team are permanent employees in an IT department and are therefore viewed as overheads.

In addition to delivery time and cost, the **nature** of **deliverables** needs to be monitored, in terms of both the size or **scope** of the deliverables and their **quality**. One danger is that the amount of functionality to be delivered increases because new requirements are discovered. If these additions to the work are not monitored and controlled, costs and delivery time will be affected. A further danger is that a task may appear to be completed, when in fact the poor quality of the resulting product means that the activity has to be re-opened to correct defects.

Thus time, cost and the scope of deliverables need to be balanced. For example, it may be possible to accelerate the progress of a late project by employing more staff, but this would increase the project cost. On the other hand, it may be possible to meet the deadline within the budgeted cost by reducing features in the application to be delivered – see Section 1.7.3, where timeboxing was described.

3.3.2 How should we monitor?

It is important to use both formal and informal methods of project monitoring. **Formal monitoring methods** include the use of written reports, email and progress meetings. The frequency, format and content of these communications should be laid down at the start of a project in a project initiation document or its equivalent.

The advantage of formal monitoring methods is that routines can be established so that people periodically focus on progress and commit themselves in writing. The disadvantage is that preparing reports can be seen as an unproductive overhead. Staff need to be convinced of the value of reporting. For example, the use of **timesheets** can be effective in establishing the staff effort expended on distinct aspects of projects, but staff need to be persuaded to fill them in conscientiously.

Many phrases can describe **informal monitoring**: keeping one's ear to the ground, management by walking about, open door policy. All of these indicate that the manager has an awareness of what team members are experiencing. Project managers need ways of maintaining good informal lines of communication with all project staff. This enables progress and problems to be communicated more quickly than with more formal methods. However, a pitfall to avoid is the alienation of team members by over-supervision.

3.4 APPLYING CONTROL

There is no point in monitoring without **control**. This is done through the **reporting cycle**. What normally happens is that the monitoring processes described above identify some shortfalls in the progress of the project. To remedy shortfalls, control needs to be applied to the project to bring it back on course. For example, staff might be transferred from non-critical work to critical activities that have fallen behind.

The reporting cycle defined in the project initiation document identifies who should be producing progress reports, with what frequency and to whom they are sent. Remember that reporting is an overhead. Reports should therefore be concise, contain only relevant information and be circulated only to those who need them. However, more concise reports require greater effort by the writer in order to save the time of the readers. As someone once apologised: 'I am sorry this report is so long: I didn't have time to write a shorter one.'

The **reporting structure** refers to the people involved in a project at different management levels. Generally progress reporting starts at the level at which work is actually done and progresses up through a hierarchy. In an IT context this means that team leaders gather progress information from their team and report up to their project manager, who then reports to the group that has been entrusted with overall responsibility for the project (the project board, project management board or steering committee). This group would include representatives of the managers of the development team, the users and the project sponsor. They, not the project manager, would have the authority to change the objectives of the project. For example, they could allocate more resources to the project or reduce the scope of what is going to be delivered. The report to the project board or steering committee is sometimes referred to as a **highlight report**. The intervals at which the reports are produced and the topics they report on need to conform to the requirements of the recipients and the importance of the information conveyed. It is important to obtain formal agreement with the reporting procedures from all the parties involved.

3.5 PURPOSE AND TYPES OF REPORTING MEETINGS

3.5.1 Team meetings

These are usually attended by team members, the team leader and possibly the project manager. A weekly frequency is usually appropriate. (In some agile projects there may even be daily 'stand-up' meetings). A report from the team leader to the project manager will be prepared. A typical agenda would include the following:

- each individual team member's progress against their plans;

- reasons for variances;

- expected progress – which looks forward to what each team member is going to do;

- current problems or **issues**;

- possible future problems – this may involve reviewing the risks that have been recorded in the project risk register (see Chapter 7) which could affect this part of the project.

It is important that all those attending have a reason for attending and a contribution to make. These meetings are often referred to as **checkpoint** meetings and the progress report produced in this case is a **checkpoint report**. (In an agile project, a **backlog** list identifying tasks completed and those to still to be done would be updated.) As issues are identified they may be recorded in an **issues log**, which will be updated as they are resolved.

3.5.2 Project board meetings

It will be recalled from Chapter 1 that project boards may have different names in different organisations – for example, steering committees or project management boards. These meetings will be attended by board members and the project manager, with secretarial support perhaps provided by a project support office. The structure and responsibilities of project boards, and the project support office, are covered in Chapter 8.

The frequency of meetings is typically monthly but the exact timing depends on the project size: larger projects may have fewer and less frequent top-level meetings, but more meetings of managers at intermediate levels. Meetings can be timed to coincide with significant project events such as the completion of a particular project phase or stage or other external triggers such as requirements for financial approvals.

Items for the agenda are similar to those for team meetings. A report from the project manager to the board is written and circulated prior to the meeting. This **highlight report** is condensed from the checkpoint reports produced in the preceding team meetings. The board is authorised to decide upon any necessary corrective action arising from progress information. This is fed back down the reporting chain and thus completes the reporting cycle.

The highlight report typically includes the following information:

- details of the progress of the project against the plan;
- current milestones achieved;
- deliverables completed;
- resource usage;
- reasons for any deviation from the plan;
- new issues and unresolved issues;
- changes to risk assessments;
- plans for the next period and products to be delivered;
- graphical representations of progress information.

3.5.3 Programme board/steering committee meetings

Organisations sometimes group projects into **programmes**, where a number of projects all contribute to a set of over-arching objectives (see Chapter 8). In these cases a programme board may be set up, to which individual project boards would report. These boards would have less frequent, less detailed meetings related to programme management, but essentially a similar agenda to those of the project boards. These would have more of a business focus than a project focus.

3.6 TAKING CORRECTIVE ACTION

Here we will examine how corrective action can be applied in a controlled way. The project manager's role is to manage on a day-to-day basis, applying minor corrections as required. However, corrections of a more major nature will need to be referred to higher authority.

This is the reason for allocating a **tolerance** within which the project manager has authority to make changes or apply corrective action. For example, you, as a project manager, may be allocated 10 per cent tolerance on time and cost on a project worth £100,000 and lasting 25 weeks. This means you are authorised to agree to changes worth £10,000 or an overrun of two and a half weeks (see Chapter 4 on change control). If a situation occurred in which the project was expected to overrun by more than two and a half weeks, this would be an **exception**. In this case, you would need to prepare an **exception report** for submission to a special meeting of the project board. This exception report would probably outline a number of **options** designed to correct the overrun and the board would decide how to proceed.

Tolerance and **contingency** pools are sometimes distinguished. Tolerances can be assigned to individual activities within the project. The contingency pool is a set of resources that is controlled by the project manager and can be allocated at the discretion of the project manager where additional resources are needed. However, if the contingency is used to buy resources in one place, less will be available for other emergencies. These resources may be augmented where activities are completed early and free up resources. Where activities can be completed early, the opportunity should be seized as this will release staff to deal with unanticipated delays elsewhere.

An exception report typically includes the following information:

- background;
- reasons why the exception arose;
- options;
- risks;
- **exception plans** showing how the project plans need to be amended in order to implement the suggested options;
- amended business case;
- recommendations.

When the project board (or equivalent) members scrutinise the exception report, they will be particularly concerned to ensure that the business case for the project will be preserved: that is, that the costs of the project will not exceed the value of its eventual benefits. If the board are satisfied with the exception report, the project manager is given authority to proceed using the chosen option and its associated exception plan. Where monitoring reveals a shortfall in the expected progress, control is applied to bring the project back on track. There are a number of stand-ard control strategies, which may or may not require an exception report. These are considered below.

Work harder, longer or faster

This is the most obvious approach and often appeals to more aggressive managers. However, although it may work to solve a short-term problem or to meet a critical deadline, it will fail if used too much. Staff will become tired, stressed and then demotivated. If overtime is paid, then project costs will increase, but not by as much as if the next option is used.

Increase resources

This needs to be carefully considered, particularly in software development. Resources in this context mean people; adding more does not usually increase productivity and often decreases it. The introduction of more staff involves a period of induction while they familiarise themselves with the work. The current staff will inevitably be involved in this process and the overall effect could even be to delay progress. An exception report and plan would be needed if the additional costs would go beyond the tolerances that have been agreed for the project.

Replan

Although some project activities have consumed more staff effort or taken longer than planned, others may have taken less effort or time. Internal movement of staff may therefore be possible, and this may be achieved without extra cost. Some may unkindly attribute this to poor planning, but the truth is that there will always be uncertainty about exactly how long tasks will take.

Extend the time scale

This is a frequent choice, but changes to deadlines will need negotiation, usually through the exception reporting process described above. Extending the deadline is often seen as weak management or as allowing the project to get out of control, but can be the most sensible option. It will, however, often increase costs as staff are allocated to the project for longer. However, sometimes the reason a project is late is that it has not been possible to assign all the staff originally planned, and so budget may not be a problem.

Reduce the project scope

This is an attractive option which also needs negotiation with management and drafting of an exception plan. Deliverables may be removed from the plan or delayed until after the planned project end date. This does not affect the originally planned cost or duration of the project, but the value of benefits to the user may be reduced. As noted in Chapter 1, this is the preferred solution of some agile project management approaches.

Terminate the project

If no acceptable alternative can be found, this may be the only remaining sensible action. Terminating the project would be justified if it is clear that the remaining costs of the project will exceed the projected value of its benefits when delivered. Despite this, terminating the project may be politically unacceptable.

ACTIVITY 3.2

In the Canal Dreams ebooking enhancement project, Activity G, *Write software* (see Figure 2.11), has been outsourced to an external software development company, XYZ. XYZ find that the task *Code provisional booking* is going to take five weeks rather than four. Their contract with Canal Dreams states that they will have to pay a penalty of £500 for each week of delay in delivering the software.

The options considered by the company are:

a) Be a week late and pay the penalty.

b) Split the 'Code provisional booking function' into two subcomponents requiring three weeks of work each and bring in an extra developer in to work in parallel with the one currently assigned to this function. Software developers cost £400 a week.

c) Get Canal Dreams to accept a delivered system on time but with some functionality missing. The supplier will agree to provide updates to implement the missing functionality at no extra charge, although this will require an additional week of coding work at a later date.

Which would be the most cost effective option for the software supplier?

3.7 GRAPHICAL REPRESENTATION OF PROGRESS INFORMATION

3.7.1 Gantt chart

Progress information can be shown on a Gantt chart by putting a bar through an activity box showing the estimated percentage completion. Figure 3.1 shows the Gantt chart that was produced in Chapter 2, updated to show the actual situation at the end of week 3 of the project. At this point the following information could be reported:

Activity reference	Name	Progress
A	Allocate room	Started a week late, completed a week later
B	Recruit staff	Reported as 50% complete, but should be more like 75% at this point. An extra day is added to planned duration
C	Analyse business requirements	Completed on time
E	Design interface	Completed on time
F	Draft acceptance test cases	Completed on time
G	Write software	10% completed
H	Devise usability tests	Start delayed by one week

We need to compare the current situation with the original plan. To do this, the details on the Gantt chart are **baselined** – that is, a snapshot is taken of the schedule at a key point. There can be several of these, but an important baseline will be the final agreed schedule at the start of the project. This is shown in Figure 3.1 by the black boxes: for instance, this shows that Task A should have started in week 1, while it actually started in week 2.

ACTIVITY 3.3

In week 4, the following actions take place:

(a) Activity B Recruit staff. There have been difficulties with finding qualified staff and effectively no progress has been made. It is reported that two further weeks, in addition to those already scheduled, will be needed.

(b) Activity G Write software. There is a discrepancy in the requirements which means that progress has been held up for the week. Currently it is hoped that the developers will be able to catch up over the next few weeks.

(c) Activity H Devise usability tests. This has now been completed.

Update the Gantt chart in Figure 3.1 to take account of these changes.

Figure 3.1 A Gantt chart that has been updated with actual progress

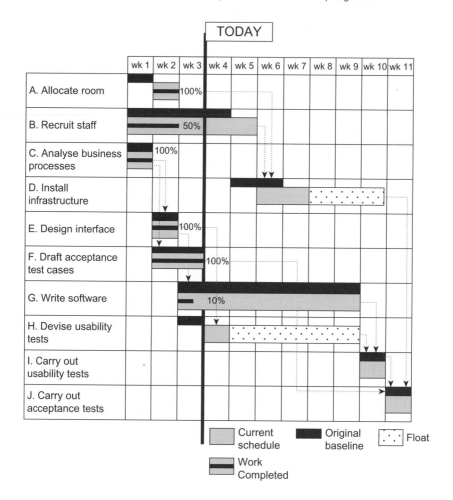

3.7.2 Cumulative resource chart

A cumulative resource chart can be used to present resource usage details (see Figure 3.2). This is sometimes called an S-curve chart, because the pattern of activity is that a project starts with a relatively low number of staff at the planning and requirements gathering stage. As the project progresses, more and more staff are needed as the development and implementation activities multiply. The demand for staff then decreases as the implementation proceeds – for example, software developers are not needed full-time once they have constructed their software components. This pattern of activity – a low demand for staff time which gradually builds up and then declines – leads to a cumulative resource chart where the line seems to have an approximate S-shape.

These charts normally have two sets of data points: one showing the expenditure that was planned and the other the actual expenditure, for comparison.

Figure 3.2 A cumulative resource chart

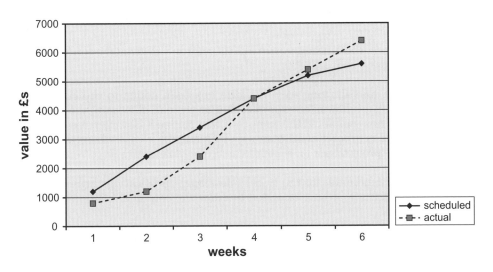

This is a convenient visual representation of project progress suitable to show to management.

Figure 3.2 shows that for most of the project we were under-spending, but there has recently been a surge in expenditure, so that currently we have spent more than planned. However, we do not know whether this is due to poor productivity, or whether we have actually produced more than was scheduled – work may have been completed early, leading to some expenditure also being incurred earlier.

The traditional S-curve chart does not show any of the following:

- whether the project is ahead or behind schedule;
- whether the project is getting value for money;
- whether problems are over or just beginning.

3.7.3 Earned value analysis

If we plot a third line, the earned value, then we can see if we are ahead or behind time, and above or below budget. **Earned value analysis (EVA)** shows the budget that was originally allocated to items of work that have been completed. When the work is finished, we can say that this value has been 'earned'. If an external supplier is involved and had a contract with fixed prices to be paid on delivery of various products, they would see these payments as earned value.

As an example, Activity G, *Write software*, is made up of a number of tasks that can be seen in Figure 2.11. To complete the overall activity within seven weeks, the supplier needs to have completed all the design work by the end of the second week of activity and all coding by the end of the sixth week. If the designers are priced at £400 a week, then at the end of the second week, four staff weeks of design work should have been completed. This means there is a **planned value** of £400 x 4 – that is, £1,600 – at the end of the second week. Now if, for example, the *Design provisional booking function* is **not** completed, because that was originally planned as two weeks' work, the **earned value** at this point in time would be only £400 x 2 for completed work – that is, £800.

In Figure 3.3, a line showing earned value has been added to the cumulative resource graph. This shows, for example, that at the end of week 6, the project is on schedule and has completed the work that was planned. However, it can be seen that expenditure is greater than planned. This may be a case where the project manager got the project back on schedule by buying in overtime.

Please note that candidates for the BCS Professional Certification (formerly ISEB) Foundation Certificate are not expected to know the details of how earned value is calculated.

Figure 3.3 Earned value graph

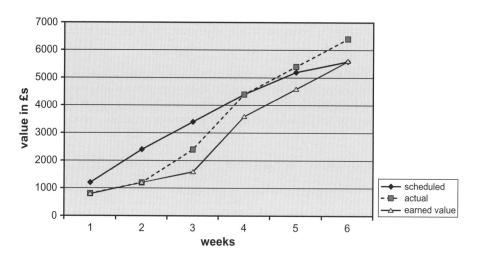

SAMPLE QUESTIONS

1. Which of the following would you most expect to see in a routine report from a project manager to a project board?
(a) Costs and benefits
(b) Progress against plan
(c) Configuration status information
(d) Current activity

2. Which of the following is not involved in collecting progress information?
(a) Team progress meetings
(b) Timesheets
(c) Comparing planned and actual costs
(d) Informal monitoring

3. Which of the following would be most likely to give rise to an exception report?
(a) A new issue being raised
(b) A proposal to make a change to a deliverable
(c) The unexpected loss of a key team member
(d) Project tolerance being exceeded

4. What is the purpose of earned value analysis?
(a) Assessing progress
(b) Collecting progress information
(c) Estimating the required effort
(d) Calculating benefits

ANSWERS TO SAMPLE QUESTIONS

1. (b) 2. (c) 3. (d) 4 (a)

POINTERS FOR ACTIVITIES

ACTIVITY 3.1

(a) If we keep to the original planned installation rate of one boatyard a day, 17 days (that is, about three to four weeks) are needed to complete installation, as there are 17 boatyards left. However, if we decide that the experience of the first week shows that the original installation rate was unrealistic, then we may project that three boatyards can be visited in a week. This would lead to between five and six weeks being needed to complete installation.

(b)(i) In this scenario, the reason for only three boatyards being visited was simply a late start rather than a low installation rate. The remaining estimate of 17 days would seem to be reasonable.

(ii) The reason for lateness here was a lower installation rate. However, as the installation programme proceeds, the installation rate should improve as the nearer boatyards are dealt with, and journey times get shorter. Revising the planned time to five to six weeks would be premature at this point.

This activity should illustrate how informal information gathering can help interpret more formal reports and the risk of extra learning time being needed if a new developer is added to the team who is unfamiliar with the application.

ACTIVITY 3.2

a) Option: be a week late and accept the penalty. The penalty would be £500, but there would also be the cost of an extra week's work (£400). This would cost £400 + £500 in all – that is, £900.

b) Option: split the functionality into two components that can be developed in parallel. Currently £1,600 (4 x £400) has been allocated to the task. The new plan would increase this to £2,400 (6 x £400) – that is, an overall increase of £800.

c) Option: staggering delivery. There would be an extra week's work for the delayed enhancement. This would cost £400.

Thus option c) would appear to be the best. Note that this is a very simplified scenario, and does not take into account many issues such as the risk of possible loss of reputation with some of the options.

ACTIVITY 3.3

See Figure 3.4.

Figure 3.4 A Gantt chart that has been updated with actual progress up to week 4

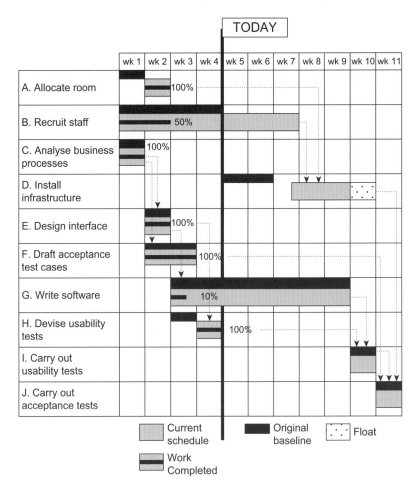

4 CHANGE CONTROL AND MANAGEMENT

LEARNING OUTCOMES

When you have completed this chapter you should be able to demonstrate an understanding of the following:

- *the reasons for change control and configuration management;*
- *change control procedures:*
    - o *the role of the change control board;*
    - o *the generation, evaluation and authorisation of change requests;*
- *configuration management:*
    - o *purpose and procedures;*
    - o *the identification of configuration items;*
    - o *product baselines;*
    - o *the content and use of configuration management databases.*

4.1 INTRODUCTION

Change management is a major concern to organisations: business changes have many implications outside the narrow confines of IT development, including their impact on staffing levels and on the skills and responsibilities required of employees. The particular focus in this chapter, however, is on the many events during a project that result in alterations to the planned work. The project manager's skill lies in controlling those changes so that minimal disruption is caused to the planned objectives of time, cost and quality. Change management ensures modifications to any aspect of the project are only accepted after a formal review of their impact upon the project as a whole.

The procedures for managing change should be established at the beginning of the project. Organisations which have kept responsibility for the maintenance of IT systems rather than outsourcing this will need some way to keep track of changes to their systems as they evolve over time in response to user and business needs. The Canal Dreams ebooking enhancement project will create new software components, change some existing software components and expand IT infrastructure: the effect of this on existing systems needs to be tracked. As well as the changes to the existing system, the planned functionality to be implemented in the ebooking enhancement may be modified during the project as the needs of the users and the business are clarified. A change control system needs to be devised and implemented, especially if outside contractors are writing the software.

Allowances need to be made within the project plans for the possibility of additional work caused by requested changes. These allowances could be part of the tolerances delegated to the project manager (see Chapter 3). The project manager needs to ensure that the effect of any change on the plan does not exceed these allowances. Where the allowance for changes is exceeded, a new project plan (the exception plan mentioned in Section 3.6) would need to be drafted and approved by the steering committee or project board.

Change management, at some level, should be applied to all changes, whether they arise from a change to user requirements or are due to design modifications. It requires participation by the users and the developers, guided by the project manager. The user must decide whether a change is essential, desirable or optional. The project manager must identify the cost, time and quality impact, and assess the feasibility of the change. Once account has been taken of users' and developers' advice, if it is decided to go ahead with the change, it is the responsibility of the project manager to implement the change.

4.2 DEFINITION OF CHANGE

In Chapters 2 and 3, we introduced the idea of the project as a sequence of activities, each of which take certain inputs and use them to produce outputs. The outputs from one process may be inputs to other processes. For example, a specification could be an input to a process that creates a system design (that is, the format of the user interfaces with the system) which will then be implemented as code. During the process, the proposed interface designs may change rapidly as the designers try out different designs and the users give feedback on them. This does not need a formal change process. At some point, however, a decision needs to be taken that the design is now in a satisfactory state to be used as a basis for constructing the system. At this point the design will need to be frozen; in other words, the design is **baselined**. If changes are made to the application after it has been baselined, then rigorous change control is needed because of the potential impact on subsequent phases.

The idea of baselines was introduced in Chapter 3 with regard to project plans. A **baselined plan** may be regarded as an agreed plan against which variations will be measured. Any change can also be seen as a potential change to a baselined plan. For projects, the baselines include:

- the agreed scope and quality of work;
- the agreed schedule;
- the agreed cost.

As noted in Chapter 3, these will tend to be interdependent. For example, if the scope of the work is expanded, then the cost will also have to be increased. Cost or scheduled duration could be reduced by decreasing the functionality in the IT application that is to be delivered.

Variations on these baselines can be categorised as follows:

- **Changes of scope**

  This type of variation generally originates outside the project, usually by the user changing the requirements or by the cost or time constraints of the project changing. In the case of the Canal Dreams ebooking enhancement scenario the users may, for example, add a requirement that when booking a boat online the client should also be able to purchase holiday insurance. On the other hand, a reduction in the project budget may mean that some system features that were planned must be dropped.

- **Development changes**

  This type of variation originates within the project and includes changes which are routinely carried out as part of the normal process of developing and refining a product. Typically, this can be something as simple as an adjustment to a screen layout.

- **Faults**

  This type of variation also originates within the project and is caused by the project team making an error. For example, errors in coding may result in erroneous results being produced.

Some discretion will be exercised in accepting or rejecting scope and development changes but changes due to design errors will normally be obligatory, since the system may not work satisfactorily unless they are corrected.

ACTIVITY 4.1

In order to encourage the UK canal holiday industry, the government decides that VAT on canal holiday bookings will be zero-rated with immediate effect. The management of Canal Dreams calculate that this could increase the demand for bookings by 30%. At the same time, the government introduces a special tax on canal holiday insurance which has to be accounted for separately on invoices. When considering the implications of changes, the project team realise that although holiday insurance was included in the original requirement, it has been missed out from the system design. What effect might these changes have on the project?

4.3 CHANGE MANAGEMENT ROLES AND RESPONSIBILITIES

It is important to clarify the various roles and responsibilities within change management. We are going to use a very specific set of terms here to identify and explain the various roles and responsibilities. In a real project environment, it is very unlikely that these precise terms will be used. However, there should be people who carry out the following roles, whatever they may be called.

- The **project manager** oversees the process and ensures that all **requests for change** (RFCs) are handled appropriately. In most cases, the project manager also has the role of change manager.

  Project managers must ensure that user representatives agree to any changes made to the project requirements. They also control the scope of the system to be developed: additional features will require more effort and increase costs. When the project sponsor agrees to additional features, adjustments may need to be made to the contractual price for the project. The project manager also needs to be on the lookout for informal changes made outside the process. Informal changes are often discovered during project reviews and a retrospective RFC form may need to be completed so that records remain accurate.

- The **change requestor** recognises a need for a change to the project and formally communicates this requirement to the change manager, by completing the RFC form.

- The **change manager** (likely to be the project manager) is responsible for logging RFCs in the **change register**. The change manager also decides whether or not a feasibility study is required for the change. Where this is desirable, for example because the extent of the impact of the change is uncertain, one or more people will be assigned to carry out the study.

- The **change feasibility group** appointed above investigates the feasibility of a proposed change. It is responsible for researching how the change may be implemented and assessing the costs, benefits and impact of each option. Its findings are documented in a feasibility study report. If the staff who are carrying out the feasibility study are also project team members, then there is a risk that project activities may be delayed.

- The **change control board** (CCB) decides whether to accept or reject the RFC forwarded by the change manager. The CCB is responsible for:
  - reviewing all RFCs forwarded by the change manager;
  - approving or rejecting each RFC based on its merits;
  - resolving change conflict, where several changes overlap;
  - resolving problems that may arise from any change;
  - approving the change implementation timetable.

- The **change implementation group** carries out the change. It is usually made up of staff from the project team.

4.4 THE CHANGE MANAGEMENT PROCESS

We have already outlined some of the processes involved in implementing a change:

- submission and receipt of change requests;
- review and logging of change requests;
- determination of the feasibility of change requests;
- approval of change requests (or rejection or putting on hold);
- implementation and closure of change requests.

It is likely that there will be a change register which is used to track the progress of an RFC through the change management process.

4.4.1 Submit request for change

In the Canal Dreams ebooking enhancement project, the project manager has been allocated the role of change manager. The office manager of the current bookings call centre is selected to act as change requestor. Requests for change can come from anyone but are all passed to the change requestor, who, in consultation with colleagues, decides whether the requested change is desirable from the users' point of view. If they decide that there is a genuine need, a request for a change is submitted to the change manager. The RFC provides a summary of the change required, including:

- a description of the change needed;
- the reasons for change, including the business implications;
- the benefits of change;
- supporting documentation.

4.4.2 Review request for change

The change manager reviews the RFC and decides the nature and scope of the feasibility study/impact analysis required for the CCB to assess the full impact of the change. In the Canal Dreams ebooking enhancement project, one request was to add the sale of holiday insurance to the online booking system. On receiving the RFC, the change manager sees that it is difficult to assess the scope of the change as some of the details of the requirement are still unclear. The impact of the change on the existing design also needs to be carefully considered. The change manager opens an RFC entry in the change register and records that a feasibility study is required.

4.4.3 Assess feasibility of change

Once the change has been logged, an assessment must be made of its feasibility and impact upon the project in terms of time, cost and quality. For small changes, a team member may assess the impact of the change in a relatively informal manner. For major changes, the feasibility study could involve several people and last for some time. Different change options will be investigated and reported on. The change feasibility study will culminate in definition of the:

- change requirements;
- change options;
- costs and benefits;
- risks and issues;
- change impact;
- recommendations and plan.

A quality review of the feasibility study is then performed in order to ensure that it has been conducted as requested and the final deliverable is approved, ready for release to the CCB. All change documentation is then collated by the change manager and submitted to the CCB for final review.

The feasibility study itself carries a cost and sometimes the project manager records these costs in the change register. These costs may well increase the budget. For external client projects, the feasibility study may be an additional service which has to be paid for by the customer.

In the Canal Dreams ebooking enhancement project, two staff are assigned to look at the requested change for a facility to purchase holiday insurance online. One is a business analyst who should understand the nature of the business and the other is an experienced developer who will be able to judge the technical impact of the requested change. After discussing the matter with the user representatives they find that the change requires two additional input fields on the booking screen, two additional data items on one of the tables in the database and some minor changes to some of the printed outputs from the system. They estimate that the changes will require three additional staff days of effort.

4.4.4 Approve request for change
A project may have a range of levels at which changes can be reviewed and approved:

- Team leaders may be allowed to accept changes that will not require additional resources and which do not affect other baselined products.

- The project manager may be allowed to decide upon changes which have a minor impact on project objectives within a tolerance level which has been agreed with the steering committee or project board.

- The CCB decides upon changes which will have a larger impact upon project objectives, but are constrained by any limitations on the budget available for changes.

- Some changes may be particularly large, but have compelling reasons for their adoption. These changes may need resources not envisaged in the original plans. In these cases, an exception report will need to be produced for consideration by the steering committee/project board.

Whatever the level of review, the change needs to be recorded and reported.

The CCB will do one of the following:

- reject the change;
- request more information related to the change;
- approve the change as requested;
- approve the change subject to specified conditions;
- put the change on hold.

The CCB will need to take account of the overall profile of possible changes. A large number of minor changes could have an overall effect that is out of all proportion to their individual significance. Approved changes will necessitate revisions to the schedule and cost plan. The CCB should prioritise changes so that those which are essential are carried out first, while non-essential changes are delayed until they can be made with the least impact on work schedules.

In the Canal Dreams ebooking enhancement project, the holiday insurance change is only one of several changes that have been requested. The CCB has a budget of only 10 staff days left to allocate, but has requests that need a total of 15 days. The CCB may accept changes requiring up to six days' effort, including the change to incorporate the holiday insurance requirements.

4.4.5 Implement request for change
This involves the complete implementation of the change, including any additional testing that may be required. On completion, the change will be signed off in the change register. Where the additional work has been carried out by an external supplier, additional invoices for the additional work will be raised by the supplier. Additional payments would not be made, however, where the change was to correct an error made by the supplier.

4.5 CONFIGURATION MANAGEMENT

The change control system described above is needed to ensure that an endless sequence of changes does not undermine the business case for the project – for example, by increasing costs so that they exceed the value of the benefits that the project will produce or by extending the time needed for development and implementation and thus reducing the benefits. Once a change has been agreed, there are further problems, such as ensuring that all documents and other products of the project are modified to reflect the change.

ACTIVITY 4.2

What deliverables of a project may be affected by the change to the Canal Dreams specification to allow for holiday insurance to be recorded when a customer books a boat online?

A project has a wealth of documents that relate to each phase of the project, along with software objects such as database structures and code segments. A very basic need is therefore for a central **project repository** or **library** where master copies of all documents and software objects are stored. Another basic requirement is that there should be a system of **version numbers** for all products so that successive baselines can be identified. These requirements mean that it is essential for one or more people involved in the project to take up a role variously called **project** or **configuration librarian** or **configuration manager**. Part of that role is to make sure that all project products are controlled, so that, for example, we can make sure that all software developers working on components that exchange information are working from the latest specifications for those components.

Configuration management has three major elements:

- configuration item identification;
- configuration status accounting;
- configuration control.

4.5.1 Configuration item identification

The items which will be subject to the **configuration management system** (CMS) need to be identified. Typically these are baselined specifications, design documents, software components, operational and support documentation, and key planning documents such as schedules and budgets. Other items, such as IT equipment, may also be subject to configuration management. These items will be defined as **configuration items** (CIs) and their details will be recorded in a **configuration management database** (CMDB). Among the details recorded in the CMDB for a configuration item are:

- a CI reference number;
- its current status;
- its version number;
- any larger configurations of which it is a part;
- any components that it has;
- other products that it is derived from;
- other products that are derived from it.

4.5.2 Configuration status accounting

After a change to a CI has been agreed, the project librarian sets the status of the CI accordingly. This process is called configuration status accounting and it maintains a continuous record of the status of the individual items which make up the system.

4.5.3 Configuration control

Configuration control ensures that due account is taken of the status of each CI. For example, when recording the change to add holiday insurance to the boat booking transaction, the configuration librarian may access the CMDB to see the current status of the software component. The librarian may find that the software component is already booked out to a software developer who is implementing a different approved change to the module. If the librarian were to release a copy of the baselined code to a second developer to add holiday insurance, that would create two different versions of the same software. The new change may be given to the developer already working on the component or there may need to be a delay while the first change is completed.

When a developer is happy that all the work associated with a change is complete, the new version of the software is passed to the librarian. The librarian then records the CI as having a new version number and as being ready for acceptance testing. Acceptance testing is usually carried out in a separate IT environment to operational processing. If the designated user representative approves the revised version of the system, a request can be made to the librarian to make the revised application operational. The former version of the software is retained in case there is a need to 'fall back' to it, if the new version turns out to be problematic.

SAMPLE QUESTIONS

1. The change control board should consist of:
(a) representatives of the key stakeholders in the project
(b) the project manager and team leader
(c) the people on the project board
(d) the project support office

2. If there are doubts about the projected costs of a proposed change, it would be the responsibility of which of the following to investigate?
(a) the change requestor
(b) the change manager
(c) the change feasibility group
(d) the change control board

3. As a documented procedure, what is the purpose of configuration management?
(a) to ensure the project remains within budget
(b) to identify and document functional characteristics of a system
(c) to record and report changes and their implementation status
(d) to verify conformance with requirements

ANSWERS TO SAMPLE QUESTIONS

1. (a) 2. (c) 3. (c)

POINTERS FOR ACTIVITIES

ACTIVITY 4.1

The changes may have the following effects on the project:

- The change to the rate of VAT should **not** involve changing the application, as there should already be a system function that allows VAT rates to be changed.
- The potential increase in bookings would not change the functional requirements, but would probably change the quality or 'non-functional' requirements. For example, the database would need to be able to hold more records. The equipment needed to run the application may need to be upgraded (this is a change to the scope of the project).
- The statutory change to accommodate holiday insurance tax could well mean changes to input screens and report layouts (this is a change to the scope of the project).
- The design's not including functions to deal with holiday insurance is a straightforward fault and the project team must correct it.

ACTIVITY 4.2

Among the many deliverables that may be affected are:

- the interface design – screens and report layouts;
- software components;
- the database structure;
- test data and expected results;
- user manuals;
- training materials.

5 QUALITY

LEARNING OUTCOMES

When you have completed this chapter you should be able to demonstrate an understanding of the following:

- *definitions of 'quality';*
- *quality control and quality assurance;*
- *measurement of quality;*
- *detection of defects during the project life cycle;*
- *quality procedures: entry, process and exit requirements;*
- *defect removal processes, including testing and reviews;*
- *types of testing (including unit, integration, user acceptance and regression testing);*
- *the inspection process and peer reviews;*
- *the principles of ISO 9001:2008 quality management systems;*
- *evaluation of suppliers.*

5.1 INTRODUCTION

The key quality concern for IT projects is providing customers with the systems they need and which meet their requirements at a price they can afford. In order to achieve this, quality needs to be built into a product. It cannot be added to a product after it has been created – except with great difficulty and cost. It is like trying to alter the foundations of a building once it is complete. To build in quality requires a commitment from all parties to the project, from the project sponsor through all the levels of management to the technical staff, customers, users and the clerical support staff. In 1979, Philip Crosby wrote a book which opened:

> *Quality is free. It's not a gift, but it is free. What costs money are the unquality things – all the actions that involve not doing the job right in the first place.*

> (Philip B. Crosby, *Quality is Free*, Mentor, 1980).

Crosby was not arguing that increasing the quality of a product did not cost money; rather, that the costs of remedying lapses in quality would be even greater. There is, as explained in Chapter 1, a relationship between quality, cost and time. The level of quality required will have an impact on both time and cost, but the more effort goes into quality at the beginning of a project, the less expenditure is needed on correcting faults at the end of the project.

5.2 DEFINITIONS OF QUALITY

A dictionary definition of quality is *'a degree or level of excellence'*, as in the phrase 'high-quality goods'. We hope all the products of a project are of a high quality. But the definition is subjective: for example, when comparing cars, people do not agree on the quality of different makes. Another definition is 'conformance to standard'. Within a project process there will be certain standards to which those developing the system ought to conform. However, the generally accepted definition which should apply to all projects is that the deliverables should be 'fit for purpose'. Again referring to cars, a Rolls Royce may not be the best vehicle if you need to get to work through heavily congested traffic on time: a scooter might be more effective. The original international standard on quality, ISO 8402:1994, formally defined quality as *'the totality of features and characteristics of a product or service which bear on its ability to satisfy stated or implied needs'.*

One aspect of this 'fit for purpose' definition – reliability – shows how the concept applies. If the Canal Dreams ebooking enhancement fails after delivery it would be annoying but not life threatening. However, if the control systems in an aircraft in flight fail, that would be disastrous. The effort, and hence the cost, of making sure that the aircraft system does not fail would be considerably greater than that required for the Canal Dreams system. The costs of a failure in required quality would also be higher. Hence the required quality varies depending on the type of system under development and the money the customer is prepared to pay. Largely, it is the customer who decides the level of quality to be built into a system. Those responsible technically for a project must advise the customer on the benefits of a well-engineered system, but finally it is the person paying who should call the tune. However, suppliers have a professional and legal commitment to the general public to ensure that the systems they produce are safe.

As software systems become more complex, it is impossible to ensure that the software will never fail. Hardware and infrastructure might also be subject to failure. Thus it is important to examine the ways in which the systems under development will behave in the event of various types of failure. This will directly influence the development process and how quality is measured. For example, a control system for a nuclear power plant will be designed to 'fail safe' – that is, in the event of a systems failure it will revert to a safe state, for instance by closing down. Obviously such a requirement must be specified in the quality criteria for the system. Where something is inherently dangerous, as in this example of the nuclear power plant, the developers would have a duty of care, not just to the project sponsor, but to the world at large.

5.3 QUALITY CHARACTERISTICS

The definition of 'fitness for purpose' needs elaborating so that it can be applied practically. Another international standard, ISO 9126-1:2001, seeks to define a set of standard characteristics by which software quality can be measured. It specifies six high-level quality characteristics:

- **Functionality:** does the system as delivered meet the functional requirements of the user? Meeting user expectations is more than just meeting a specification.

- **Reliability:** how often does the system break down and how long does it take to put right? Are the results it produces consistent?

- **Usability:** is it straightforward to use? How much training is required for someone to be able to use it?

- **Efficiency:** what level of physical resources are required to operate the system?

- **Maintainability:** all software is subject to change; can this software application be modified easily and without introducing unexpected errors?

- **Portability:** how easy is it to take the software from the particular platform on which it was developed to another environment?

The standard itself goes into a lot more detail for each heading, but these top-level qualities are useful guides, particularly when trying to establish measurable quality criteria. The qualities can be subdivided into sub-qualities which are then measured. For example, in the Canal Dreams ebooking enhancement, response time in answering queries about the availability of boats is an important part of usability. For the application, engineering measurements have to be mapped to a value on a scale reflecting user satisfaction – for example, a response time under five seconds might correspond to 'acceptable'.

5.4 QUALITY CRITERIA

To ensure that quality is built into a product the level of quality required has to be specified at the beginning, before the product is developed. In Chapter 2, the concept of a **product definition** was introduced. Each product definition includes a section headed **quality criteria**. The accurate completion of this section enables project team staff to check the product is fit for purpose later on in the development cycle, by seeing whether it meets the quality criteria specified.

In order to achieve this, the criteria themselves have to have three qualities:

- They must be **specific**.

- They must be **measurable**.

- They must be **achievable**.

As an example, the aircraft control system could have a specific requirement that the product must never fail. That is quite clear. It is also measurable: by testing the product for an exhaustive amount of time until the product fails, it can be demonstrated that the requirement has not been met. In this case, we can prove when the quality criterion has **not** been met, but we cannot prove that it will not fail at some point in the future. Hence the 'never' requirement is not achievable. We can, however, provide an estimate of the probability of failure which will get smaller as testing continues. As noted with the nuclear power station, we can also plan for the possibility of failure, for example, by allowing a reversion to manual control.

An organisation should develop standards for product definitions or other documents which contain quality criteria as the basis for its quality procedures.

ACTIVITY 5.1

Refer back to the discussion about product definitions (Section 2.2.1). Draw up quality criteria that can be used to assess a product definition document (not the product it defines). Add any other headings that you consider should be standard for all product definitions. Specify how the criteria can be measured.

ACTIVITY 5.2

The following are examples of good and bad quality criteria:

All screen layouts should have similar layouts and use the same terminology.
Screens should be user friendly.
The system should be able to handle 50 transactions.
The system should allow for 20 users at any one time without degradation.
The response time should not be longer than three seconds.

Comment on the effectiveness of each of these quality criteria.

ACTIVITY 5.3

Maintainability is defined as a quality characteristic. How can it be measured?

ACTIVITY 5.4

How can the reliability of a system be measured?

5.5 QUALITY CONTROL VERSUS QUALITY ASSURANCE

Having introduced the concept of quality criteria and how they should be speci-fied, we now discuss how the quality criteria of a product created by a project are checked. In an ideal environment the project will take place in an organisation committed to quality and with standards already in place for certain activities. If this is not so, part of project set-up will be the creation of the framework for manag-ing the quality of the project.

The quality framework is called the **quality management system** (QMS). It may be based on the ISO 9001 series of standards (see Section 5.10). Within the QMS, there will be a quality strategy and quality assurance processes. They have to be reviewed and, if necessary, modified to meet the specific requirements of each new project.

A **quality strategy** defines the QMS and includes:

- procedures and standards for creating a **project quality plan**;
- a definition of **quality criteria**;
- quality **control** procedures;
- quality **assurance** procedures;
- a statement of compliance with or allowed deviation from **industry standards**;
- **acceptance criteria;**
- an allocation of responsibilities for defining or undertaking quality-related activities.

The methods for exercising quality control are discussed in detail later, but gener-ally these tend to take the form of a review. Quality assurance stands alongside reviews but is external to them. Quality assurance is like an audit. It is designed to confirm that proper procedures are in place and have been applied correctly. The difference between the two can be illustrated by the following example. A metal rod must be 15mm in diameter ($\pm$ 0.1mm). Control is achieved by passing each rod through a measuring device which triggers rejection if the rod is out of tolerance. Assurance is exercised by checking that the measuring device is accurate and that all rods go through it.

For each component, quality criteria are specified. As we have seen above, this is in effect the first part of quality control, for without the criteria no control can take place. A control process is then undertaken to ensure that the criteria have been met. This is followed, at an appropriate time, by an assurance process which confirms that the agreed procedures have been followed and that all products have undergone the necessary checks.

You will recall from Chapter 1 that the systems development life cycle has a number of stages – for example, requirements analysis is followed by systems design, followed by construction and testing. Each stage contains several quality control processes, in the form of some sort of review, and a quality assurance process takes place at the end of each stage. If proper quality control is lacking, corrective action may be mandated. The quality control and, if necessary, the quality assurance processes are repeated until the quality criteria are met.

5.6 QUALITY PLAN

The production of a **quality plan** is vital to the success of a project. It specifies the particular standards that apply to the project. They could be existing standards – for example, laying down the content and layout of product definitions. Modifications to existing standards may be needed because of the special characteristics of a new project. In some cases they could be completely new, as a project goes into territory not previously explored. The standards may derive from industry standards or be developed internally. If there are any gaps, then the project manager must ensure that the appropriate standards are specified.

The plan also specifies how, when and by whom the quality control activities should be undertaken, the quality assurance processes to be followed and who will carry them out. It may also include configuration management and change control procedures (see Chapter 4).

The quality plan itself is subject to quality control and quality assurance processes. Quality control would normally be an integral part of the project team's work. The quality assurance activities, on the other hand, are usually carried out by people outside the project team who report directly to the steering committee/project board. This separation of responsibilities helps to ensure that the process is transparent and reduces possible conflicts of interest.

5.7 DETECTING DEFECTS

5.7.1 The V Model

The V Model is a useful model of the systems development process (see Figure 5.1) in which the solid lines represent the forward progress of the project and the dotted lines represent the way in which quality control is exercised.

There are two quality control processes at work: one between stages and the other across the V. For example, the requirements specification describes all the functions and quality attributes that the customer requires in the system. This should include an acceptance test plan showing how the requirements are going to be assessed in the final system. Using the acceptance test plan, user acceptance testing can be undertaken to demonstrate that the final system to be delivered does indeed meet the requirements identified. This link between the requirements specification and user acceptance testing is shown by the dotted arrow between the two, identified as the 'acceptance test plan'.

Systems design follows requirements specification and is joined to it by an arrow. As part of the quality control process for systems design it is necessary to check that everything specified in the requirements specification has been incorporated

Figure 5.1 A simplified V model

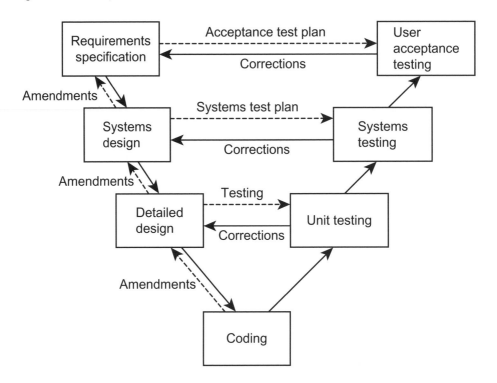

into the design. If not, then the design is incomplete. It is also important to ensure that no requirements have been introduced into the design that were not present in the original requirements. Systems design has two dotted arrows: one across to systems testing and the other back to requirements specification. The horizontal arrow shows the systems testing that needs to be carried out to validate the design after it has been implemented. The arrow to the requirements specification indicates that the design process may discover errors in the systems requirements. For example, gaps may be found in the definition of the requirements, or two requirements may be found to be inconsistent. The same processes are applied to each stage of development.

5.7.2 Process requirements
The importance of quality criteria for each product was stressed earlier. Within a project, as we saw in Chapter 2, a product is the output from a process or activity. The qualities that are required in a product are captured by following the correct processes that create the product. This makes it necessary to specify for each stage, and within each stage for each activity, the **process requirements**:

Entry requirements state what must exist before the stage or activity can begin. For example, before design can begin, the requirements documentation must be agreed and signed off and any design techniques to be used must be specified. If new techniques are being introduced, then the necessary training should have been given. If design begins before requirements are agreed, this will lead to problems if the requirements are changed while the design is underway. Either the design will have to be reworked, at additional cost, or the need to amend the design may be forgotten and only picked up much later, when making changes will be more expensive.

Implementation requirements define how the process should be done. For example, in design, implementation requirements may specify the use of techniques such as entity modelling or logical process modelling. Implementation requirements also specify the software tools that are to be used.

Exit requirements indicate what should be in place for a successful sign-off of the activity. The exit requirements for design documentation are that it:

- is complete;
- has been reviewed;
- meets the standards agreed;
- covers all the requirements for this component;
- leaves no outstanding issues.

5.7.3 Defect removal process

Before a defect can be removed, it must be identified. This is relatively straightforward – though more costly – in the later stages of development, when test cases can be run though the system to see if they give the expected results (dynamic testing). In the earlier stages of a project, different techniques have to be applied, such as:

- desk checking;
- document review;
- peer review;
- inspection;
- walkthrough;
- pair programming;
- static testing.

When specifying quality criteria that apply to, say, a software specification, the technique for testing if those criteria are met should be specified, along with the type of staff who will implement the technique and the tools they will need.

Desk checking

Desk checking is the most basic of the review activities. Authors, or creators, of products review what they have produced. This may mean reading it through and checking for errors in the case of a document, or working through the logic of a program to identify mistakes. It will, hopefully, remove many trivial errors before the product is subjected to more vigorous scrutiny.

ACTIVITY 5.5

Read through the following table and identify any errors:

Data	Item range	Cross-checks
Time: hours	0–23	
Time: minutes	0–60	
Time: seconds	0–60	
Day	1–31	Cannot be greater than 30 if month = 2, 4, 6, 7, 9 or 11
Month	1–12	
Year	>2010	

Document review

In a **document review**, one or more people other than the author read a document to ensure that it meets the specified quality criteria. For example, is it complete, unambiguous, self-consistent and consistent with related documents? Is it clear? Are all technical terms properly used and understood? Does it conform to the agreed layout? If the document is a requirement, then such a review could well be carried out by the potential user of the system.

Peer review

A **peer review** is a particular type of document review that is often carried out on a design document or actual code. The author's co-workers (or **peers**) examine copies of the document and make comments about it. The issues considered include:

- Is the proposed design technically feasible?

- Is there an easier or better way to achieve the same objectives?

- Does the design conform to company standards for such processes?

- Can the design communicate with other parts of the system?

- Does the design cover all the requirements that should be included in this part?

- Are there any ambiguities?

The peer reviewers of software often **dry-run** the code – that is, take some test input data and manipulate it on paper according to the instructions in the code.

Peer reviews can be done relatively informally within the project team. However, the time needed by the reviewers needs to be officially scheduled – after all, they have their own software on which to work as well.

Inspection

An **inspection** is a more formalised way of reviewing a product. Its purpose is to review the product in order to identify defects in a planned, independent, controlled and documented manner. It is a process with the following structure:

- Preparation:

  o setting up the review meeting, including the time, place and who is to attend;

  o distributing documentation – for example, the product and its description;

  o reviewers annotating the product, if it is a document, and recording defects.

- Meeting:

  o discussion of potential defects identified by reviewers which should confirm: whether they are defects or not, but not seek to produce a solution;

  o agreement of follow-up appropriate to each defect.

- Recording:

  o follow-up actions and responsibilities;

  o agreement of outcome and sign-off if appropriate.

- Follow-up:

  o advising the project manager of the outcome;

  o planning remedial work;

  o signing off when complete.

There are four roles within this process:

(1) The **moderator** sets the agenda and controls the review process, ensures actions and required results are agreed and, once the process is complete, signs off the review.

(2) The **author** provides the reviewers with relevant product documentation, answers questions about the product during the review and agrees actions to resolve defects.

(3) **Reviewers** undertake the review, assessing the product against the quality criteria, identifying potential defects and ensuring that the nature of any defect is clearly understood by the author.

(4) The **scribe** takes notes of the agreed actions, who is to carry them out and who is to check corrections. He or she confirms these details at the end of the meeting.

Walkthrough

A **walkthrough** is a particular technique which can be used during an inspection or on its own. The author takes the audience through the documents and they feed back their comments on it. The audience may be drawn from both technical and user groups, each of which would have a different view on the documents. For example, IT infrastructure management staff may be concerned about the proposed system's impact on what they already manage. Again there would be a scribe to record any actions agreed.

With peer reviews, inspections and walkthroughs, no attempt should be made during the meeting to solve the problems identified. Problems should be recorded; the author will then go away and seek to come up with a solution. The review can then be repeated if it is felt that the changes required were of sufficient significance. An alternative is for one person to be instructed to ensure that the necessary alterations have been made.

Pair programming

The review techniques described above depend on copies of documents, including code, being printed and additional reviewing activities being scheduled. To avoid this, in agile development environments, code developers sometimes work in pairs. The pair take turns to type in code at the workstation while the other advises and checks on what is being entered. This is rather like a real-time version of a peer review.

Static testing

Some software tools (programs) carry out **static testing** by analysing the structure of the code. Such tools look at the branches and loops in a program and calculate a measure of complexity. The more complex a software component is, the more difficult it will be to maintain. In addition, dynamic analysis can identify code which is not executed by the test data used. This is useful as it may show up a shortcoming in the test data or unneeded code in the software.

All the above processes take place during the activities on the left-hand side of the V model. The quality control processes on the right-hand side are dominated by dynamic testing.

5.8 DYNAMIC TESTING

Dynamic testing is divided into various levels:

- unit;
- integration;
- systems;
- user acceptance;
- regression.

For each type of testing, a set of test data and a set of expected results must be produced. Referring back to the V model, a test plan should have been produced at the appropriate stage in the development process on the left-hand side of the V. The plan should contain the necessary guidance for producing the test data, if not the test data itself.

Unit testing

Unit testing is the very basic testing to be carried out on a piece of software. Test data for unit testing should be designed to cover the usual range of input expected for the system. Each function that the software is expected to handle should be tested. The test should take place not just once but in various combinations and different sequences. Often problems lie in the combinations of data and transactions. Testing is then extended to cover, for example, numbers just inside and just outside any specified limits. Alphabetic fields are tested with entries longer than that which the system should permit. Mandatory fields are left blank. This is not an exhaustive list but is indicative of what is necessary. All tests are designed to ensure that this particular unit will not fail because of bad data or unusual combinations but will handle them in a predefined way.

ACTIVITY 5.6

Assume that the table of time/date checks drawn up (and corrected) in Activity 5.5 has now been implemented in an input screen in an IT system. Draw up a set of test data and expected results that could be used to test that the checks on data are being carried out correctly.

Each set of test results should be carefully checked against the expected results to ensure agreement. Any discrepancies should be noted and reported so that the software can be amended and retested. Sometimes, however, the expected results are wrong! It is important to keep records of all faults found and resulting changes made, so that if problems arise at a later stage there is a trail which can be followed to establish how they were introduced.

There are tools – commonly called **test harnesses** – which can simulate programs or subroutines that supply data to or use data from the module under test. The test harness can record data input and output and the routes through the program which have been exercised. Automated tools can also simulate keyboard input and capture and compare screen output with expected results.

Once unit testing has been successfully completed the unit can be signed off and registered as a configuration item (see Chapter 4).

Integration testing

Integration testing links a number of system components and runs them as a whole. This checks that the units communicate properly with each other. The sorts of things that come to light at this stage include:

- data items which are treated as being in different formats in different units;
- reference codes given different meanings by different authors;
- conditions set by one component that another cannot cope with.

Some of these can be avoided by the use of shared databases and database management systems (DBMS), but transaction files which link different sub-systems can still be mistakenly defined to have different formats in different places.

As errors are found they will be recorded and corrected. The integration test will then need to be repeated.

Systems testing

Systems testing is the final stage in testing by the development professionals. It involves running the whole system on the infrastructure that will be used when the system is operational. It may sound just like a step up from integration testing but there are many additional issues which must be addressed, for example:

- Does the system run on the infrastructure to be used for the final system?

- Are the response times within the tolerances set in the requirements specification?

- Can the system cope with the planned workloads?

- What is the effect of high loading on the system?

Again, there are tools to help. They can be used to simulate large numbers of users and high volumes of data.

User acceptance testing

User acceptance testing is the crucial test. Can the users operate the system? Does it meet their expectations, not just their requirements? Users should be involved in the development process from the beginning of the project and should have had sight of how the system is working before this point is reached so it now contains few surprises. Expectations as well as requirements have to be managed. It may or may not be helpful to involve users in earlier testing activities: it is a matter of keeping a balance between helping them to understand the way the system works, and them seeing all the faults that arise during testing and becoming disillusioned.

Acceptance testing underlines the importance of having clear requirements from the start. Users should have a well-defined acceptance test to guide them as they test whether the system meets their expressed needs. Where discrepancies are found, the fault must be recorded, the source of the problem established and the system reworked and retested.

Regression testing

Regression testing is quite different from the earlier testing processes. Regression testing needs to take place at each stage of testing. Whenever a fault is found and the offending piece of software identified, it has to be corrected. Unfortunately this may well introduce further errors or uncover ones that were masked by the first error. Regression testing involves running an agreed set of test data through the system again to confirm that the original error has been corrected and no further errors have been introduced or uncovered. The later a fault is discovered in the testing hierarchy, the more regression testing has to take place.

Regression testing is also needed whenever the system is changed, whether during development or after implementation, as changes made to one part of a system can adversely affect another. Regression testing can largely be automated by using a standard set of test data and expected results.

5.9 EVALUATING SUPPLIERS

It is usual these days for systems to be developed by third parties. It is therefore important to establish whether those third party suppliers have the necessary quality procedures in place to ensure that the software to be supplied is to the standard expected.

To do this one has to investigate the supplier, ask some searching questions, examine documentation and perhaps undertake auditing of their processes. The types of questions to be asked include.

- Do they operate to ISO 9001 (the international standard for quality management systems)?

- How are their projects organised? (For example, do they follow PRINCE2, the standard for project management procedures sponsored by the UK government?)

- Do they use a defined development methodology, such as the Dynamic System Development Method (DSDM Atern)?

- How is quality control exercised?

- At what points are quality reviews held?

- Is there a quality assurance process?

- Is there a configuration management system in place?

- How are change requests handled?

This list is by no means exhaustive. The response to each question should be supported by evidence. For example, if it is claimed that software quality is assessed by reviews, then examples of the outputs from the reviews can be examined. Observing some of their reviews can increase confidence in their quality processes.

It must be recognised that the supplier and the customer have different business objectives. Making a project a success therefore needs both parties to see the project as a joint venture. The customer cannot simply hand it over to the supplier and wait for delivery, nor should the supplier hide the development process from the customer. Each side needs to be involved to an appropriate level for the type of project being undertaken.

5.10 ISO 9001:2008

ISO 9001:2008 is the international standard for quality management and is specifically aimed at producers and suppliers of any products and services, not necessarily software. Organisations are inspected and awarded ISO 9001 certification by accredited auditors. This means that as a potential client you can assume a particular standard of quality management without having to carry out detailed checks yourself.

However, while ISO 9001 states that a quality level should be specified, it does not say what that level should be. For example, when producing a ballpoint pen it could be stated that a quality requirement is for the ink in the pen to last three years, but equally it could state three days. If the requirement is met then the expected quality is achieved even if it is very low. Thus having a set of ISO 9001 procedures only guarantees that a level of quality has been specified, not that this level is accepted universally as being appropriate. This does allow the client of a ISO 9001:2008 supplier to negotiate the quality criteria they personally need.

ISO 9001:2008 is based on the following principles:

- **Customer focus:** understanding and meeting or exceeding the customer requirements;

- **Leadership:** providing this for the organisation to give it the purpose, unity and direction to achieve quality objectives;

- **People:** involvement of staff at all levels of the organisations involved;

- **Process approach:** attention to individual processes which produce intermediate or deliverable products;

- **Systems approach to management:** focusing on inter-related processes producing deliverable products;

- **Continual improvement;**

- **Factual** approach to decision-making;

- **Mutually beneficial** client-supplier relationships.

The TickIT Guide is designed to show the application of ISO 9001:2008 to software systems. In doing this, it also takes account of another ISO standard, ISO/IEC 90003:2004, which offers guidance on the application of ISO 9001:2008 to software engineering. The current (2012) version of the TickIT Guide, Issue 5.5, is in line with ISO 9001:2008. TickIT applies whenever software development is carried out, the software is incorporated in the delivered product or service and the organisation wishes to be ISO 9001 accredited. TickIT is managed and maintained by the British Standards Institution (BSI).

ADVANCED TOPIC
Capability maturity models

5.11 CAPABILITY MATURITY MODELS

In the discussion of ISO9001:2008, we assumed that the motivation was to allow customers to assess potential suppliers. Sometimes, however, the managers of a supplier organisation want to assess their own quality processes to find ways of improving them. They want to gauge their current level of effectiveness, but also identify what needs to be done to get it to a higher level. This brings in the concept of **maturity modes,** where the organisation is assessed as being at a particular level of **process maturity**.

On example of this is the capability maturity model (CMM) originally developed by the Software Engineering Institute at Carnegie Mellon University for the US Department of Defense. This comprises five levels.

(1) **Initial.** Any organisation would be at this level by default. Good quality work may be done, but customers cannot be sure that this is always the case.

(2) **Managed.** Some basic project management and other systems are in place.

(3) **Defined.** The way each task in software development is done is defined to enable consistent good practice.

(4) **Quantifiably managed.** Processes and their products are measured and controlled – for example, the number of errors created in each process.

(5) **Optimising.** The measurement data collected is analysed to find ways of improving processes.

The assessment of an organisation's maturity level can be done internally, or external auditors could be employed so that the maturity level can be published externally, as in the case of ISO 9001.

SAMPLE QUESTIONS

1. 'Fitness for purpose' defines which of the following?
(a) The quality of the project deliverables
(b) The usability of the delivered IT application
(c) The capability of the staff who will implement the IT application
(d) The capability of the staff who will use the system when it is operational

2. User acceptance testing is an example of which of the following?
(a) Project control
(b) Project monitoring
(c) Quality control
(d) Quality assurance

3. Which of the following is NOT a defined role in inspections?
(a) The moderator
(b) The project manager
(c) The scribe
(d) The author

4. ISO 9001:2008 is a standard that defines which of the following?
(a) Project management standards
(b) IT project deliverables
(c) Quality management systems
(d) Software testing procedures

ANSWERS TO SAMPLE QUESTIONS

1. (a) 2. (c) 3. (b) 4. (c)

POINTERS FOR ACTIVITIES

ACTIVITY 5.1

Quality criteria could include:

(1) A product description must contain the following headings or sections (from Chapter 2):

- identity;
- description;
- derived from;
- components;
- format;
- quality criteria.

Additional headings could include the following:

- author;
- owner;
- date first compiled;
- date of last amendment;
- version number.

(2) The 'derived from' section must refer to valid product types.

(3) The 'format' section must provide enough information to allow someone to create an instance of the product in the correct form.

One can easily identify other criteria.

These criteria meet the requirement of being specific (for example, a list of contents), measurable (the sections are either there or not) and achievable (it is possible, without difficulty, to ensure that each heading is present).

The quality of a product description will most likely be measured through a review process, such as inspection by a fellow developer. What this process does not do is to ensure fully that the correct information is entered for each heading. This may require a further set of quality criteria which again should be subject to review.

ACTIVITY 5.2

Criterion	Assessment
All screen layouts should have similar layouts and use the same terminology	This is relatively clear and measurements could be devised. But more detailed checklists of things to look for (as might be found in a style guide) would be helpful.
Screens should be user friendly	This is subjective and therefore not measurable.
The system should be able to handle 50 transactions	This is not measurable as there is no indication as to the period of time within which the transactions have to be handled.
The system should allow for 20 users at any one time without degradation	This is better than the previous criterion, but clarification of what each user might be doing could be requested.
The response time should not be longer than three seconds	This is a mixture. It is clear that a response time of three seconds or less is required but it does not specify under what conditions.

Ideally the last two criteria should be combined so that a baseline of three seconds is given with a loading of 20 simultaneous users. This can still be improved upon. For example, it could be stated that the response time should be less than four seconds for at least 95 per cent of the time with a loading of 20 simultaneous users and should never exceed 10 seconds. Such a statement does allow for the odd occasion when the three seconds might be exceeded. These examples show how difficult, yet important, it is to get the quality criteria correctly specified for each product in the development process.

ACTIVITY 5.3

As will be seen later in the main text of this chapter, there is a difference between measuring the quality of an existing software component – where actual performance can be measured – and assessing the likely quality of an application as it is being built. In an existing software component, we could collect statistics about the amount of effort that has been needed to implement actual changes.

If the software is being created, we could examine the code to see if it has characteristics that are likely to lead to maintainability. A system would be maintainable if it satisfied the following criteria (among others):

- The structure of the software is clear.
- The names used for items of data and procedures are indicative of what is being referred to.

● Code is clear and unambiguous.

● Documentation is present to support code.

There are other possible software engineering criteria that could be discussed, such as loose coupling of components (minimal cross-references) and cohesion (all code for a function being together).

The measurement for these criteria would probably be a peer review process.

ACTIVITY 5.4

Two ways in which reliability of a system has been traditionally measured:

(1) Meantime between failure (MTBF) specifies how long the system runs without failing. These days, this would be specified in weeks or even months. Something which fails every day would not be very popular with those operating the system.

(2) Meantime to repair (MTTR) specifies how long it takes to repair the system when it fails. It is no good if a system cannot be restored to a working situation in a reasonable length of time. That length of time needs to be specified as part of the acceptance criteria. Note that this measure is also related to the attribute of maintainability.

Other valid measurements could be considered, such as the percentage availability of the system.

ACTIVITY 5.5

The errors include:

Time: minutes should be in the range 0–59.
Time: seconds should be in the range 0–59.
Day: July (month 7) has 31 days, February never has more than 29 days and there is no leap year check. The cross-check should be 'If month = 2 and year is leap, up to 29; if month = 2 and year is not leap, up to 28; if month = 4, 6, 9 or 11, up to 30'.

ACTIVITY 5.6

Data item	Test description	Input	Expected result
Day	Not numeric Outside lower range Inside lower range Outside upper range	XX 0 1 32	Error message Error message Accept Error message
Month	Not numeric Outside lower range Inside lower range Outside upper range Inside upper range	July 0 1 13 12	Error message Error message Accept Error message Accept
	Cross-check with day	day = 31 month = 1, 3, 5, 7, 8, 10, 12	Accept
	Cross-check with day	day = 31 month = 2, 4, 6, 9, 11	Error message
	Cross-check with day	day = 30 month = 4, 6, 9, 11	Accept
	Cross-check with day	day = 30 month = 2	Error message
Year	Not numeric Outside lower range Inside lower range	xx 2009 2011	Error message Error message Accept
	Cross-check with day	day = 29 month = 2 year divisible by 4	Accept
	Cross-check with day	day = 29 month = 2 year not divisible by 4	Error message

You may find some 'holes' in the above test data. It should illustrate that although devising test data is not the most glamorous job, creating effective test cases does require the kind of attention to detail that we normally expect of software developers. Devising test data will also trigger questions about the precise nature of the requirements – for example, is there really no upper limit on year?

6 ESTIMATING

LEARNING OUTCOMES

When you have completed this chapter you should be able to demonstrate an understanding of the following:

- *the effects of over- and under-estimating;*
- *effort versus duration;*
- *the relationship between effort and cost;*
- *estimates and targets;*
- *use of expert judgement, including its advantages and disadvantages;*
- *the Delphi approach;*
- *top-down estimating*
- *bottom-up estimating;*
- *the use of analogy in estimating.*

6.1 INTRODUCTION

In Chapter 2, we explained how to draw up a plan for a project. One of the things that we did was to allocate an estimated duration to each of the activities to be carried out. This allowed us to calculate the overall duration of the project and to identify when we would need to call upon the services of individuals to carry out their tasks. In this chapter, we will explore further the ways in which these estimates can be produced.

6.2 WHAT WE ESTIMATE AND WHY IT IS IMPORTANT

6.2.1 Effort versus duration

As well as estimating the time from the start to the end of an activity, it is also necessary to assess the amount of effort needed. Duration should not be confused with **effort**. For example, if it takes one worker two hours to clear a car park of snow then, all other things being equal, it takes two workers only one hour. In both cases, the effort is two hours but the activity **duration** is two hours in one case and only one hour in the other. There can be cases where the duration is longer: for example, where someone only works in the afternoons on a particular task. In fact, a problem is that activities often take longer than planned even though the effort has not increased. This may happen, for instance, when you have to wait for approval from a higher level of management before a job is signed off. This distinction between effort and duration can be particularly important when assessing the probable cost of a project, as on some projects staff costs are governed by the hours actually worked (typically where staff complete timesheets), while on others the costs are governed by the time in which people are employed on the project (even if there is not always work for them to do).

6.2.2 The effects of over- and under-estimating

If effort and duration are under-estimated, the project can fail because it has exceeded its budget or has been delayed beyond its agreed completion date. This may be so even when staff have worked efficiently and conscientiously. Allocating less time and money than is really needed can also affect the quality of the final project deliverables: team members may work hard to meet deadlines but, as a consequence, produce sub-standard work.

On the other hand, estimates that are too generous can also be a problem. If the estimate is the basis for a bid to carry out some work for an external customer, then an excessively high estimate may lead to the work being lost to a competitor. Parkinson's Law ('work expands to fill the time available') means that an excessively generous estimate may lead to lower productivity. If a task is allocated four weeks when it really needs only three, there is a chance that, with the pressure removed, staff will take the planned time.

6.2.3 Estimates and targets

Identifying the exact time it will take to do something is very difficult because, if the same task is repeated a number of times, each instance of the task execution is likely to have a slightly different duration. Take going to work by car. It is unlikely that on any two days this will take exactly the same amount of time. The journey time will vary because of factors such as weather conditions and the pressure of traffic. This means that an estimate of effort or time is really a most likely effort/time with a range of possibilities on each side of it. Within this range of times we can choose a **target** – we can go for an 'aggressive' target which may get the job done quickly, but with a strong possibility of failure, or a more generous estimate which is likely to expand the length of time needed, but have a safer chance of the target being met. The target, if at all reasonable, can become a **self-fulfilling prophecy** – with the commuting example, if you know that you are going to be late you may take steps to speed up, perhaps by taking an alternative route if the normal one is congested. Estimating can thus have a 'political' aspect. Some managers may reduce estimates, either to gain acceptance for a proposed project, or as a means of pressurising developers to work harder. There are clearly risks involved in such an approach, as well as possible ethical issues.

6.3 EXPERT JUDGEMENT

6.3.1 Using expert judgement

Where do you start if you want to produce reasonable estimates? Although estimating is treated as a separate, isolated topic in project management and information systems development, it in fact depends on the completion of other tasks that provide information for estimates. For a start, you need to know:

- What activities are going to be carried out during the course of the project;
- How much work is going to have to be carried out by these activities.

For example, to work out how long it will take to install some software on all the workstations in an organisation, we need to know approximately how long it takes to install the software on a single workstation and how many workstations there are in the organisation. We may also need to know how geographically dispersed the workstations are. The best person to tell us about these things would be someone familiar with the tasks to be carried out and the environment in which they are done. As a general rule, the best people to estimate effort are those who are experts in the area. As a consequence, most guides to estimating identify **expert opinion** or **expert judgement** as an estimating method.

Although 'phoning a friend' can be a very sensible approach, there remains the question of how the experts themselves derive their estimates. There is a possibility that they have their own experts upon whom they can call, but at some point someone must sit down and work out the estimate based on their own judgement – and the likelihood is that they will end up using the analogy approach described below.

The advantages of using expert judgement include the following:

- It involves the people with the best experience of similar work in the past and the best knowledge of the work environment;
- The people who are most likely to be doing the work are involved with the estimating process – they will be more motivated to meet the targets set if they have had a hand in setting them in the first place.

There are, however, some balancing risks:

- The task to be carried out may be a new one of which there is no prior experience;
- Experts can be prone to human error – they may, for example, underestimate the time that they would need to carry out a task in case a larger figure suggests that they are less capable;
- It can be difficult for the project planner to evaluate the quality of an estimate that is essentially someone else's guess;
- Large, complex tasks may require the expertise of several different specialists.

6.3.2 The Delphi approach

One method that attempts to improve the quality of expert judgement is the **Delphi technique** which originated in the Rand corporation in the USA. There are different versions of this, but the general principle is that a group of experts are asked to produce, individually and without consulting others, an estimate supported by some kind of rationale. These are all forwarded to a moderator who collates the replies and circulates them back to the group as a whole. Each member of the group can now read the anonymous estimates and supporting rationales of the other group members. They may now submit a revised opinion. Hopefully, the opinions of the experts should converge on a consensus.

The justification for the technique is that it should lead to people's views being judged on their merits and undue deference will not be paid to more senior staff or the more dominant personalities.

6.4 APPROACHES TO ESTIMATING

We are now going to discuss key approaches to estimating. However, first we are going to explain the terms **bottom-up** and **top-down**. Note that these are not specific estimating methods, but describe a way of grouping estimating methods.

6.4.1 Bottom-up

In essence, we break the task for which an estimate is to be produced into component sub-tasks and then break the component sub-tasks into sub-sub-tasks and so on, until we get to elements that we think would not take one or two people more than a week to complete. The idea is that you can realistically imagine what can be accomplished in one or two weeks in a way that would not be possible for one or two months. To get an overall estimate of the effort needed for the project, you simply add up all the effort for the component tasks.

This method is also sometimes called **analytical** or **activity-based estimating**. Some people (especially those who are or who have been software developers) find the name 'bottom-up' confusing because the first part of the process is really top-down!

ACTIVITY 6.1

Which planning product identified in Chapter 2 could be the basis for an initial bottom-up estimate?

A bottom-up estimate is recommended where you have no accurate historical records of relevant past projects to guide you. A disadvantage of the method is that it is very time-consuming as you have, in effect, to draw up a detailed plan of how the project is to be carried out first. It could be argued that you are going to have to do this anyway. However, it may be a very tedious and speculative task if you have been asked for a rough estimate at the feasibility study stage of the project proposal.

ACTIVITY 6.2

You have been asked to organise the recruitment of staff for the new network support centre needed for the Canal Dreams ebooking enhancement. Identify the component activities in this overall task, as you would for the first stage of the bottom-up approach to estimating effort.

6.4.2 Top-down

With the top-down approach, we look for some overall characteristics of the job to be done and, from these, produce a global effort estimate. This figure is nearly always based on our knowledge of past cases.

An example of top-down estimating is when house owners have to make decisions about the sum for which they should insure their house. The question here is the probable cost of rebuilding the house in the event of it being destroyed, for example by fire. Most insurance companies produce a handy set of tables where you can look up such variables as the number of storeys your house has, the number of bedrooms, the area of floor-space, the material out of which it has been constructed and the region in which it is located. For each combination of these characteristics a rebuilding cost will be suggested. The insurance company can produce such tables as it has access to many historical records of the actual cost of rebuilding houses.

This is essentially a top-down approach because only one global figure is produced. In the unhappy case of a fire actually occurring, this figure would not help a builder to calculate how much effort would be needed to dig the foundations, build the walls, put on the roof and all the other individual components of the building operation. A builder may be able to use past experience of the proportion of total costs usually consumed by each type of activity, such as foundation digging.

6.5 A PARAMETRIC APPROACH

The base estimate created when using a top-down approach can be derived in a number of ways. In estimating the costs of rebuilding a house, a **parametric** method was used. This means that the estimate was based on certain variables or parameters (for example, the number of storeys in the house and the number of bedrooms). These parameters can be said to 'drive' the size of the house to be built: you would expect a house with three storeys and five bedrooms to be physically bigger than a bungalow with only two bedrooms. These parameters are therefore sometimes called **size drivers**. You would also expect the three-storey building to need more work, or effort, to build than the bungalow. These parameters are therefore also sometimes called **effort drivers**.

6.5.1 Size drivers and productivity

Earlier we had an example where a technician was allocated the job of installing a new piece of software on every workstation in an organisation. Clearly, the more workstations there are, the bigger the job and the longer its duration. Hence the number of workstations is a size driver and an effort driver for this activity.

ACTIVITY 6.3

Identify the possible size and effort drivers in the Canal Dreams ebooking enhancement for each of the following activities:

(a) Creating training material for users;
(b) Analysing business processes;
(c) Carrying out acceptance tests;
(d) Writing and testing software.

In order to produce an estimate of effort using this method, we also need a productivity rate. For example, in addition to the number of workstations we would need to know the average time needed to install the software on a single workstation. This time per workstation would be the **productivity rate**. If this rate was 12 minutes per workstation and there were 50 workstations then we could guess the overall duration of the job would be around 50 × 12 minutes – that is, about ten hours.

The usual way to obtain the productivity rate is from records of past projects. Where these are not available within an organisation, it is sometimes possible to obtain 'industry' data that relates not to projects in a single organisation, but to projects in a particular industrial sector. This kind of information can help managers compare the productivity in their organisation with that of others – this is sometimes called **benchmarking**. If they find that they have much lower productivity, this may spur them on to search for more productive ways of working. However, caution needs to be practised if the reason for using industry data is that local project data is missing: there can be large differences in productivity between organisations, because organisations and their businesses are so different.

ACTIVITY 6.4

In the earlier example about the time needed to drive to work, identify:

(a) the size driver;
(b) the productivity rate;
(c) other factors that may cause a variation in the time it takes to get to work.

The additional factors are called **productivity drivers**. A key productivity driver when it comes to developing and implementing IT systems is experience. When putting a figure on how long a technical activity like developing software code is going to take, more experienced estimators will try to find out who will be doing the work.

Productivity drivers vary from activity to activity, but other drivers often include:

- the availability of tools to assist in the work;
- communication overheads, including the time it takes to get requirements clarified and approved;
- the stability of the environment – that is, the extent to which the work has to cope with changes to requirements or resources;
- the size of the project team: there is a tendency for larger jobs involving lots of people to be less efficient than smaller ones because more time has to be spent on management, planning and co-ordination at the expense of 'real work'.

The problems that can affect productivity are often considered at the same time as risks to the project in general (see Chapter 7).

6.5.2 Function points

There was a time when almost all IT projects involved writing software of some description. This is now increasingly less the case for many reasons, one of which is the tendency to use 'off-the-shelf' software applications. However, there are still many cases where software has to be written specially, and these situations can cause particular challenges for the estimation of development effort.

If we use a parametric approach, the first question is what to use as size drivers. If IT is old enough to have any real 'traditions', then one of the longest established of these would be to use **lines of code** as the size driver for software development. (When software is written, the programmer writes the instructions – as lines of code – in a form which is comprehensible to human experts. This 'code' is an electronic document which can be changed, added to and printed. When the code is to be executed by the computer, the document is 'read' by a special piece of software which converts it into a format that the computer can interpret automatically.)

From this very brief explanation it can be seen that:

- the code is a very technical product – it would need a software expert to estimate the number of lines of code;
- you will not know the exact number of lines of code until quite near the end of the project; most other size drivers are known at the beginning, or at least at an early stage, of the project.

Things are also complicated by there being many programming languages. Some are more 'powerful' than others – that is, they need fewer lines of code to carry out a particular procedure.

Rather than use this technical unit of size which is invisible to everyone except the software developers, it is more convenient to use as the size drivers counts externally apparent features of the software application. This would be rather like using the number of storeys, the floor space and so on to estimate the cost of a house, rather than the number of bricks. With software applications, this can be done with function points.

ADVANCED TOPIC Function points

For the purposes of the Foundation Certificate, you do not need to know the details of the rules of function point counting. There are at least two major systems of function point counting and some of the detailed rules are rather arcane, to say the least. The following description should be enough to give a general idea of the approach. (It is based on one particular version – Mark II, or Symons, function points – simply because this is, in our view, the simplest method for getting an understanding of the general principles of the approach.)

(a) The size drivers are features of the software that are apparent to the user. In general, users are aware of the **transactions** that they can carry out when using a software application. A transaction is where the user **inputs** something into the computer (normally by typing), the computer carries out a procedure and comes up with a **response**, normally in the form of an **output**, and the computer system is left in a **consistent state**. This is similar to a use case in UML.

When booking a boating holiday using the new Canal Dreams ebooking system, you make a mistake (for example, typing in an invalid date) and an error message is displayed. Although an input has been followed by an output (the error message), the system is not in a consistent state: only half the booking details have been set up. In this case the processing so far would not be regarded as a transaction: either the whole booking would be rejected or the processing would continue until a complete and correct set of booking details had been input.

(b) For each type of transaction, a count is made of the number of items of information that are input and output, and the number of tables of information that are accessed. In general, it can be assumed that the more of these there are, the more lines of code will have to be written, and the more work there will be for the system developers.

(c) The counts are weighted to take account of the relative difficulty of implementing each type of feature. For example, a simple output is normally easier to implement than an input. With inputs you often have to carry out error checking, which adds to the developer's work. To take account of these differences in difficulty, the feature counts are weighted appropriately. In the Mark II method, inputs are weighted 0.58, outputs 0.26 and entity (or table) accesses 1.66. Effectively this means that the weighting between inputs, outputs and entity accesses is about 2:1:6. The use of such peculiar numbers is because the inventor of this method wanted the resulting function point counts to be about the same as for the American method (specified by the international function point user group, IFPUG) and hoped to achieve this by making the weightings add up to 2.50 (that is, 0.58 + 0.26 + 1.66).

A restriction on the use of function point counting is that it assumes that there is a human operator initiating transactions and receiving outputs from the system. COSMIC function points are an alternative approach that can be used to measure the size of embedded software which interfaces with other software and hardware layers rather than human users.

6.5.3 An example of function point counting

Within the Canal Dreams ebooking system, there is a transaction which records the final payment made by a customer for a booking for which they have already paid a deposit. There are **three inputs** for a new payment:

- date;
- customer account reference;
- amount.

There are **four** possible **outputs** from the transaction:

- payment reference, a number allocated automatically by the computer system;
- customer name;
- customer address;
- an error message.

To carry out this transaction, a CUSTOMER ACCOUNT table and a PAYMENT table are accessed, giving **two entity accesses**. The function point count for this transaction is therefore:

$$(3 \times 0.58) + (4 \times 0.26) + (2 \times 1.66), \text{ that is, } 6.10.$$

What does this 6.10 really represent? It is best regarded as an index value that gives an idea of the amount of processing carried out by the transaction. For a single, isolated transaction, this measure is not very accurate. However, if you were able to add up the function point counts for all the transactions in an information system, then it is likely that the count for the application as a whole would be a useful indicator of its size.

We can use a function point count to find out the relative productivity of development projects that have already been completed. We may find that the average number of function points implemented per day is around five. This may seem a rather small number, but 'development effort' here includes the whole development cycle, from requirements gathering to testing. When a new project proposal comes along, a preliminary investigation may suggest that the delivered system would have a count of about 250 function points. The estimated effort is therefore in the region of 250/5 days – that is, 50 days.

6.6 ESTIMATING BY ANALOGY

The function point approach (and, indeed, the more generic approach of using size drivers and productivity rates) is based on the assumption that we have the details of the size driver values and actual effort of past projects. Often, however, such records do not exist. For smaller organisations particularly, the IT projects that have been previously implemented may all seem to have their own peculiarities. For example, some may have involved the installation of off-the-shelf packages, others may have required specially written software, some a mixture of the two, and so on. This seems to suggest that previous experience is not a stable basis for estimating the effort for new projects. However, in this kind of situation the **analogy** or **comparative** approach could be used.

The main steps with this method are as follows.

(a) Identify the key characteristics of the new project.

(b) Search for a previous project which has similar characteristics.

(c) Use the actual effort recorded for the previous project as the base estimate for the new one.

(d) Identify the key differences between the old and the new projects (it is unlikely that the old project is an exact match for the new one).

(e) Adjust the base estimate to take account of the identified differences.

An analogy approach can be used to create a top-down estimate for a project. Where there is no past project which seems to be a useful analogy for the new project, an estimator can use analogy to select parts of old projects that seem similar to components of the current project (using analogy as part of a bottom-up approach).

As Table 6.1 shows, both analogy and the parametric approaches can be used either at the overall level of a project or for estimating the effort needed for components. The activity-based approach – breaking down the overall task into smaller components – seems almost by definition to be a bottom-up approach.

Table 6.1 Relationship between top-down/bottom-up and the three main estimating approaches

	activity-based/ analytical	parametric	analogy/ comparative
top-down		✓	✓
bottom-up	✓	✓	✓

6.7 CHECKLIST

As a project planner you may often need to use the effort estimates produced by experts from technical areas in which you are not knowledgeable. Are there any ways in which you can realistically review these estimates? It may be possible to assess the plausibility of the estimates by asking the estimator the questions below.

- What methods were used to produce the estimates?

- How is the relative size of the job measured (in other words, what are the size/effort drivers)?

- How much effort was assumed would be required for each unit of the size driver (in other words, what productivity rates are you assuming)?

- Can a past project of about the same size be identified which had about the same effort?

- If a job with a comparable size cannot be identified, can past jobs which had similar productivity rates be found?

SAMPLE QUESTIONS

XYZ ORGANISATION SCENARIO

Staff have managed to develop information systems at a rate of five function points per staff day. A new system has been assessed as requiring 120 function points to implement, but the staff available are relatively inexperienced and are only 80% as productive as the staff usually used in such projects.

1. An under estimate of effort is MOST likely to lead to which of the following?
(a) decreased productivity
(b) decreased quality
(c) a less competitive bid for a contract
(d) a longer project duration

2. Which of the following estimating methods is MOST likely to be used bottom-up?
(a) parametric
(b) algorithmic
(c) Delphi
(d) activity-based

Both questions 3 and 4 use this scenario.

3. In the XYZ scenario, of which of the following is 80% of the value?
(a) a size driver
(b) an effort driver
(c) a productivity rate
(d) a productivity driver

4. In the XYZ scenario, what would be the best estimate of effort for the project?
(a) 30 days
(b) 25 days
(c) 24 days
(d) 20 days

ANSWERS TO SAMPLE QUESTIONS

1. (b) 2. (d) 3. (d) 4. (a)

POINTERS FOR ACTIVITIES

ACTIVITY 6.1

The work breakdown structure (or possibly the product breakdown structure).

ACTIVITY 6.2

Among the activities that may need to be carried out are:

- Create/agree job descriptions
- Create job advertisements
- Collect and assess applications and curricula vitae (CVs) from potential employees
- Invite selected candidates
- Interview candidates
- Notify successful and unsuccessful candidates
- Request, await and check references
- Confirm appointment
- Arrange induction
- Carry out induction processes

This set of activities offers some good illustrations of the difference between elapsed time and effort. There will be some points – for example where you are waiting for references – where little effort is expended but time will be passing.

ACTIVITY 6.3

The following are suitable answers:

(a) The number of functions that users need to be able to use.

(b) The number of different types of system user (as each will need to be interviewed for their requirements), and the number of different operations carried out in the system.

(c) The number of functions to be tested and the number of input and output data items to be tested.

(d) The number of different functions in the system, the number of inputs, outputs and tables accessed.

ACTIVITY 6.4

(a) The size driver would be the distance driven to work.

(b) The productivity rate would be the average speed of the car.

(c) We have already suggested that the weather and the amount of traffic congestion could have an effect on the travel time.

In this case, the weather and traffic do not increase the size of the job to be done – the distance to work remains the same. These factors are best seen as influences on the productivity rate. In order to assess more accurately the time it takes me to go to work, I could take account of these intermittent constraints on my speed. I may be aware, for instance, that the rush-hour traffic in the morning tends to be significantly less heavy during school holidays. I could therefore perhaps allow myself to start off to work a few minutes later when it is half-term. On the other hand, I may start earlier if the weather is foggy, as I know that this can slow down the traffic.

7 RISK

LEARNING OUTCOMES

When you have completed this chapter you should be able to demonstrate an understanding of the following:

- *identification and prioritisation of risks;*
- *assessment of the probability and impact of risks, i.e. risk exposure;*
- *risk reduction activities versus contingency actions;*
- *typical risks associated with information systems;*
- *assessment of the value of risk reduction activities;*
- *maintenance of risk registers.*

7.1 INTRODUCTION

There are a number of definitions of risk, some embracing opportunities as well as threats. Arguably, the most used is the definition provided in PRINCE2, the project management standard sponsored by the UK government:

'The chance of exposure to the adverse consequences of future events.'

Those adverse effects could be a reduction in the value delivered in the project, or even the failure of the project. They include higher development costs, delayed project completion, reduced scope and reduced performance. The adverse effect could also be a less effective completed system that fails to deliver the appropriate capability, which in turn means that the original business case is not fully realised.

In Chapter 1 we distinguished between **project objectives** and **business objectives**. Similarly, some risks are **project-related** – that is, they threaten the successful achievement of the project's objectives – and others are **business risks**. The project team may meet the project objectives and deliver the required IT application on time and within budget. Business conditions, however, may mean the client organisation cannot exploit the application to achieve their business objectives. In the Canal Dreams ebooking enhancement scenario, a slump in the demand for canal holidays could mean that the investment in the new booking system does not pay for itself. On the other hand, risks involve opportunities as well as threats. Unexpected events might be to our advantage if we can modify our plans quickly to take account of them.

The risks come from within the project, or they could be caused by external events. If an adverse event has already occurred, this is not a risk but should be treated as a **project issue** – that is, a problem that needs to be resolved.

External (or environmental) risks include:

- government intervention;
- cuts in resources, including staff;
- reduction in financial support;
- increased competition from rivals;
- social developments.

Internal risks include:

- staff changes;
- lack of policies which can guide decision-making;
- increased scope of changes;
- lack of developer experience;
- sabotage.

The objectives of risk management are to identify, address and minimise risks before they become threats to the successful completion of a project. However, we need to be aware of the 'law of diminishing returns', which suggests that the initial effort and expenditure provides the best return and that the benefits from further spending to solve a problem gradually become smaller. Buying one smoke detector for your house when you have none could make a big difference to your safety, but buying another when you already have five will probably make little difference. Actions to eliminate some risks may be prohibitively costly, difficult or lengthy and, if adopted, would adversely affect the business case.

7.2 RISK MANAGEMENT

The management of risks is similar to the management of any activity. There is a planning and control cycle similar to the one described in Chapter 3. Risk management is a continuous process throughout the life of the project. However, before managers form any plans, they need to identify the risks and make some decisions about them. A typical risk management framework is shown in Figure 7.1. Major risks have to be identified and then plans have to be made to deal with them. These plans include activities that enable other activities to be undertaken in the event of a risk turning into a real problem. For example, taking back-ups of important

Figure 7.1 Risk management framework

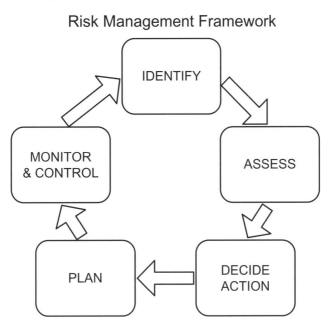

computer files allows them to be restored if the originals are damaged. The project is then executed and is monitored and controlled to see where risks have materialised and where appropriate actions need to be initiated.

7.3 IDENTIFYING RISKS

Before the risk identification process begins, there will be a number of facts or issues that are already known. These include particular views, trends or constraints within the project development environment. It is worth examining these facts as they often provide the causes of the risks that will eventually need to be managed.

ACTIVITY 7.1

Reread the Canal Dreams ebookings enhancement project scenario in Chapter 1 and identify features of the project or its environment which you think could lead to difficulties.

There are a number of aids to building the initial list of risks. Many risk textbooks, application development documents and company standards provide **prompt** or **checklists** containing a number of **generic** business and project risks that originate from outside and inside the project. These can be used to determine which risks in the list apply to the project. One well-known list of risks is Barry Boehm's 'Top Ten Software Project Risks':

(1) **Personnel shortfalls** – for example, developers not being familiar with the technologies needed for the project;

(2) **Unrealistic schedules and budgets** – in some cases this could be because not all essential requirements have been identified;

(3) **Developing the wrong functions and properties;**

(4) **Developing the wrong user interface;**

(5) **Gold-plating** – this is development of software functionality that is not really needed and ends up not being used;

(6) **Continuous stream of requirements changes;**

(7) **Shortfalls in externally furnished components;**

(8) **Shortfalls in externally performed tasks;**

(9) **Real-time performance shortfalls;**

(10) **Straining computer-science capabilities** – current technologies and expertise are not sufficiently well developed to satisfy the requirements and the project becomes effectively a research rather than a development project.

As well as **generic** risks applying to almost any project, there are **specific risks** peculiar to the circumstances of the particular project – most of the risks to the Canal Dreams ebooking enhancement project identified in Activity 7.1 were specific to that project. Methods of gathering information about specific risks include:

- Interviewing experts or stakeholders within the project;

- Brainstorming workshops with stakeholders to identify risks – the discussion of a risk by one member may lead to others recognising further risks;

- Searching past project documentation.

It is important however, in view of the law of diminishing returns, not to assume that all generic risks will be relevant. Dismiss risks not really belonging to the project. For example, an aircraft crashing into the software development laboratory is a risk, but not one usually considered unless the laboratory is under a busy flight takeoff path.

It is helpful, in identifying a risk, to recognise the true focus of the problem. For example, you should not say:

'Development of the application software may overrun.'

Instead, you should say:

'There is a risk that the application programmers are too inexperienced in the chosen language and therefore the application software may be completed late.'

Note that this risk statement is only appropriate when it is set in the context of the given project. In this example, it is a fact that the programmers are inexperienced in the chosen language. This fact, or issue, is not a risk but a cause of a potential risk. A risk has an element of uncertainty about it. The fact that there is limited experience of the chosen programming language only becomes a risk in our project if the difficulty of the design and coding of the application software makes demands that are too great for the inexperienced programmers.

ACTIVITY 7.2

Given the list of possible issues listed in Activity 7.1, identify six or seven possible risks for the Canal Dreams ebooking enhancement project.

7.4 ASSESSING THE RISK

7.4.1 Risk evaluation criteria

We recognised that we may not be able to take action for all possible risks identified. Therefore there is a need to prioritise the risks, but before this can happen there is a need to evaluate the risk itself. The evaluation is based on two key criteria:

- the **probability** that the risk will occur;

- the **impact** that the risk will have, should it occur.

Together, risk and impact give an idea of the magnitude of the risk – the **risk exposure**. A third factor is the **proximity** of the risk, which takes account of the fact that the magnitude of the risk may vary throughout the project – for example, once coding has been successfully completed some risks relating to coding disappear.

7.4.2 Risk exposure

We noted above that the **impact**, or severity, is the adverse effect that the risk will have on the project should it materialise. This impact may be a longer development **time**, a reduction in the **scope** of the deliverable, a reduction in the **performance of the deliverable**, or an increase in the **resources needed**. The scope and the performance are often combined as a reduction in **quality**. The increased resources, both of materials and labour, are usually referred to as increased **costs**. Should a risk occur, it may impact time, cost or quality, and of course any impact will affect the business case.

A risk can be viewed as an **opportunity**. Let us examine the example above, of developing application software in a chosen language (say Java) with which developers are not familiar. The plan for the software development tasks may increase the expected duration of the tasks to take account of the developers' lack of experience. If, however, the developers were able to pick up Java very quickly their tasks may be completed earlier than scheduled. In this case the project manager ought to exploit this opportunity and start the next task as soon as possible. The time gained here will be a useful buffer if problems occur later in the project.

Impact is not the only issue that affects the seriousness of a risk (or risk exposure). A risk could cause immense damage if it occurs – as in the example above of the aircraft crashing into a workplace – but in practice be dismissed because of the minute probability of it occurring.

7.4.3 Risk proximity

The **proximity** relates to the time period in the project when the risk may occur. Risk evaluation may identify that a given risk is more likely to occur during a particular activity, or that after a certain project milestone it will no longer be applicable, or that the impact could change depending on when the risk occurs. The risk of inexperienced Java programmers delaying completion of work will have an impact on the software development stage. Once the milestone marking the end of software coding has been successfully passed, this will no longer be a risk. In Chapter 1, it was noted that uncertainty about a project was greatest at the beginning because of all the unknowns associated with a new project. As knowledge is gained about the application and technical domains during the project, much of this uncertainty is reduced.

7.5 QUANTITATIVE APPROACHES TO RISK

Risk assessment can be **quantitative** – based on seemingly precise mathematical values – or **qualitative** – based on broader management intuition.

ADVANCED TOPIC

Quantitative risk assessment

When a quantitative approach is used, probability is represented as either a percentage between 0 per cent and 100 per cent or a value in the range of 0.00 to 1.00. Zero per cent or 0.0 means there is absolutely no chance of something happening, while 100 per cent or 1.00 means it is absolutely certain that something will happen. A probability of 0.40 means there is a 4 out of 10 chance of something happening.

Impact is most conveniently measured as a money value reflecting the financial loss of the risk should it actually occur, but is sometimes measured in time (that is, the amount of delay caused).

The values of probability and impact can be multiplied to create a risk exposure. For example, if there were a 0.10 probability of IT equipment worth £20,000 being stolen, that the risk exposure would be 0.10 × £20,000 – that is, £2,000. (Note that all the numbers here are picked for ease of the arithmetic, not because they are realistic.) Crudely, this risk exposure value can be compared to the amount that might be paid as an insurance premium. If there were 100 organisations with IT equipment of the same value and the same chance of theft and they all contributed £2,000 to a pool, the pool would be big enough for 10 per cent of them to withdraw £20,000 if they were robbed. (This is a simplified model: in real life the 10 per cent would have to be based on an average over several years. It is unlikely that it would be exactly 10 per cent in any one year.)

Risk exposure = impact × probability

An advantage of the quantitative model is it is easy to assess the effectiveness of a risk reduction action. Say that in the above example an organisation decided to buy a burglar alarm for £1,500 (Once again, this figure has been picked simply to make the calculation easier) and it is estimated that it would reduce the probability of a successful theft to 1 per cent (or 0.01).

A risk reduction leverage can be calculated as follows:

Risk reduction leverage (RRL) = (RE_{before} − RE_{after}) / cost of risk reduction

RE_{before} is the risk exposure before the risk reduction action is taken – that is, £2,000. RE_{after} is the risk exposure after the action (the installation of the burglar alarm) – that is, £20,000 × 0.01, or £200.

The calculation of RRL is therefore (£2,000–£200) / £1,500) = 1.2.

Because an RRL is greater than 1.0, it means that the reduction action is worthwhile. (This could be compared to the cost of the burglar alarm being offset by a reduction in insurance premiums.)

There are a few problems with the practical application of quantitative risk assessment, including the following:

- Unless you have a very large set of data about past occurrences of the particular risk, identifying the probability of a risk may end up as little more than guesswork.

- In our simplistic example, the cost of the theft was exactly £20,000. In practice the amount of damage can vary, and so this value could be guesswork. Where there is a large amount of data about past occurrences of the risk it may be possible to produce a table showing the probability of different ranges of cost – but this kind of information is unlikely to be available to a project planner.

- Quantitative risk exposure values are based on the principle that when risks actually occur, the situation can be remedied by using resources put aside to meet possible losses. It assumes that the remedies when risks occur can be partly paid for by resources put aside for other risks that have not been used. However, this assumption does not hold where the loss caused by a particularly large risk occurring is simply too large and would exhaust the fund. The bankruptcy of the client organisation might be an example of these **show-stoppers**.

7.6 THE QUALITATIVE APPROACH TO PROJECT RISK ASSESSMENT

Because of these problems, modern practice in project risk management tends to adopt a more qualitative approach, and it is on this that we will focus in the remainder of this chapter.

7.6.1 Risk probability

While quantitative risk assessment requires access to data about past projects, there is a wider range of approaches to obtaining qualitative assessments including interviewing experts or stakeholders and brainstorming in a workshop. Various qualitative descriptions of probability can be used to describe the probability of a risk occurring, such as 'extremely likely', 'very high', 'high', 'medium','low', 'very low' or 'improbable'. Similar descriptions are used in the qualitative assessment of the impact the risk will have on cost, quality or time.

Quantitative values can still be assessed but these will be expressed as being within a range – for example, a 20–50 per cent probability of occurrence – and then be mapped to one of the categories of the probability and the impact (see Tables 7.1 to 7.4).

Table 7.1 Mapping qualitative and quantitative assessments of risk probability

Index	Impact level	
4	High	Greater than 50% chance that the risk will occur
3	Significant	30–50% chance that the risk will occur
2	Moderate	10–29% chance that the risk will occur
1	Low	Less than 10% chance that the risk will occur

Table 7.2 Mapping qualitative and quantitative assessments of cost impact

Index	Impact level	
4	High	Greater than 20% above project cost tolerance
3	Significant	Up to 20% above the project cost tolerance
2	Moderate	Greater than 50% of the project cost tolerance but still within it
1	Low	Within 50% of cost tolerance

Table 7.3 Mapping qualitative and quantitative assessments of scope impact

Index	Impact level	
4	High	Inability to meet mandatory project functionality
3	Significant	Shortfalls in key functionality
2	Moderate	Shortfalls in secondary functionality
1	Low	Some minor functions missing

Table 7.4 Mapping qualitative and quantitative assessments of time impact

Index	Impact level	
4	High	Greater than 20% above project time tolerance
3	Significant	Up to 20% above the project time tolerance
2	Moderate	Greater than 50% of the project time tolerance but still within it
1	Low	Within 50% of project time tolerance

The Delphi method (see Chapter 6) is a more formal version of the expert approach to risk assessment. Risk assessment is very closely associated with effort estimation and in some cases can be carried out at the same time.

7.6.2 Prioritising risks

Once the evaluations or risk probability and impact have taken place, the risks can be prioritised to ensure that the risk management effort is placed where it is needed most. We have already seen that where a quantitative approach has been taken, a risk exposure value can be calculated. The bigger this value is, the more serious it is. However, this calculation cannot be done if we have made qualitative assessments. To aid decision making for qualitative assessment, a **probability impact grid**, sometimes known as a **summary risk profile**, can be used (see Figure 7.2). In the grid the numbers uniquely identify each risk. Some organisations will have three probability impact grids to cover the impacts on time, cost and quality respectively. Other organisations combine them. Organisations refine their understanding of the risk profiles over time and become able to judge more accurately the threat of a risk and the action to be taken to deal with it.

Figure 7.2 Probability impact grid

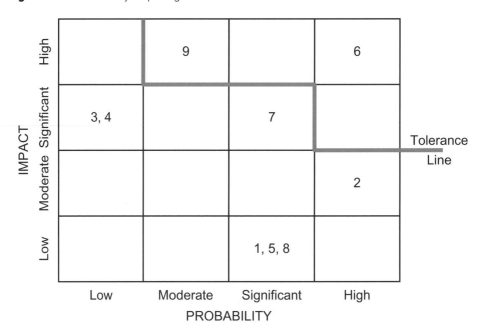

7.7 DECIDING THE APPROPRIATE ACTIONS

Uncertainty about a project due to the risks identified can be reduced by taking **mitigating action**. Mitigation is defined as an action to reduce, transfer or eliminate a risk. Any mitigating action chosen will consequently mean revisiting the project schedule, the development costs, the functional scope and the performance of the deliverables and updating them, if necessary, to take account of the chosen action. In addition to the option of adopting one or more mitigating actions for a given risk, the project team could decide to take no action at all or to develop a contingency plan.

7.7.1 Accepting the risk
The organisation could accept the risk and take no further action other than to monitor it. It could be argued that taking no action is not, strictly speaking, a form of mitigation, but it is mentioned here for the sake of completeness. This option could be chosen if the risk probability is low, the impact is acceptable, or it is thought that none of the other actions listed below would be appropriate. The actions might be rejected because the cost of action outweighs the impact cost – the risk reduction leverage measures this – or because it is not practical in the particular context. In the case of the programmers' Java inexperience, the cost of action was judged to exceed the cost of the impact of the risk.

7.7.2 Preventing the risk
This is sometimes called 'risk avoidance'. The organisation could prevent the risk from occurring by changing the activity in which the risk is embedded. In the example above, where the programmers' inexperience of the chosen language is a risk,

the risk might be prevented by deciding to develop the application software in the language with which they are familiar.

7.7.3 Reducing the risk

The organisation could still take the planned action but reduce the probability of the risk occurring or the impact of the risk. Again any reduction action will take place before the expected risk occurs. In the above example, if it was decided to reduce the possible impact on the project deadline, steps could be taken to try to ensure that the development was not late despite the inexperience of the programmers. For example, one or more specialists experienced in the chosen language could be recruited to join the development team to act as advisers to reduce technical misunderstandings of the features of the 'new' programming language.

7.7.4 Transferring the risk

The organisation may take action to transfer the risk to another party. For example, an organisation with inexperienced programmers could contract the work out to a software house. If the software house was working to a fixed-price contract then the problem of possible cost over-runs would be transferred to them.

7.7.5 Contingency

The organisation may decide not to take any action before the risk occurs, but instead to plan an action to be taken once the risk occurs, or if it becomes more certain that the risk will occur. For example, it is very difficult to think of many practical ways of eliminating the possibility of key staff becoming ill at a critical point in the project. Contingency planning might identify staff who could cover the role of a staff member made unavailable in such circumstances. This action differs from other actions as generally it only incurs costs if the risk materialises. As with all actions, there will be costs associated with managing the risk (see Section 7.5). There may also be costs associated with creating the conditions which would allow the contingency action to take place, as is the case when back-ups of files have to be taken to allow the contingency action of restoring files if they are corrupted.

ACTIVITY 7.3

Explain the actions that may be taken to mitigate the risks to the Canal Dreams ebooking enhancement project scenario that were noted as a result of Activity 7.2.

When selecting the mitigating actions to be taken, it may well be that more than one action is appropriate for a particular risk, or that a chosen action can mitigate more than one risk. For example, in the case of the Canal Dreams ebooking enhancement project, retraining some of the current telesales team to work in the new network support centre might not only reduce demotivation but also reduce the risk of not finding suitable staff to operate the new network support centre.

Because the action chosen will have an impact on the original project plans, it is important to ensure that the benefits of the action outweigh the benefits of inaction – the calculation of risk reduction leverage could contribute to this.

How many actions to approve, and in relation to which risks, are key decisions in risk management. Initial focus is likely to be on the **show-stoppers** – risks that would prevent completion of the project.

Where a quantitative approach has been adopted, the risk exposure figures for individual risks could be summed to get an overall project risk exposure. If necessary, a number of actions could be planned to reduce the **risk exposure** of the project to a level acceptable. Alternatively actions could be taken to address just the highest priority risks.

With the qualitative approach, a **risk tolerance line** has been shown on the probability impact grid in Figure 7.2. The organisation will not approve a project with risks that occur above this line. Therefore action would be taken to ensure a new position on the grid for these risks by reducing their probability or impact, or both.

7.8 PLANNING, MONITORING AND CONTROL

In the paragraphs above, it has been assumed that the risk identification, risk assessment and mitigating actions have occurred in the earlier stages of a project, during the project planning. However, all of the processes described above will continue throughout the life of the project, as new risks may be identified at any time, and there may also be secondary risks that result from actions to reduce initial risks. For example, outsourcing software development to a software house because of the inexperience of in-house developers will itself generate new risks. The monitoring of risks should be part of the project control cycle (see Chapter 3). The monitoring process is most likely to be a mixture of regular timed reviews and reviews held after significant events, particularly the end of a project stage. It is important to have a structured project risk plan to document the planning and to facilitate the monitoring and control process. This will consist of a **risk register**, also known as a **risk log**, which lists all the risks identified with the project (see Figure 7.3).

Figure 7.3 Risk register

Risk register							
Risk Id	Risk title		Risk owner	Post-risk management				PlanNo
				Prob.	Cost	Qual	Time	
1								
2								
3								
4								
5								
6								
7								
8								
9								

Typical headings are shown in the form, but others could be added – for example, a risk category or the earliest date that the risk could occur. Initially, not all entries would be completed. As risk assessment is an iterative process, entries such as the post-risk management values are entered after the appropriate actions have been approved and the risk plans formulated.

Figure 7.4 Risk record

Risk Record					
Risk Id	Risk title			Risk category	Plan No.
Risk owner		Initiation date	Earliest Occurence	Review date	Deletion date
Risk description					
Impact description					

Probability/impact values	Probability	Impact on		
		Cost	Qualty	Time
Pre-risk management				
Post-risk management				

History			
Version	Date	Action	Comments

For each risk in the risk register, an individual **risk record** will be created (see Figure 7.4). Note that Figure 7.4 shows the probability and impacts both before and after any agreed mitigating action is taken – see the lines for 'pre-risk management' and 'post-risk management'. In addition to the risk register and risk records, plans of the actions chosen need to be documented. As noted earlier, there is not necessarily a one-to-one relationship between a risk and a risk plan. An individual risk may necessitate a number of plans before the risk exposure is reduced to an acceptable level, or there may be one plan that addresses a number of identified risks.

Figures 7.3 and 7.4 refer to a **risk owner**. The risk owner is responsible for ensuring adequate management of the risk, including how and at what intervals a risk will be monitored. If the nature of a risk changes during this process, it may be necessary to revise earlier risk mitigation plans.

7.9 SUMMARY

Managing risk is a continuous process. It involves identifying the risks and analysing them to establish their probability, impact, proximity, exposure and priority. Remedial actions need to be determined and plans produced to implement these actions, followed by scheduled monitoring and appropriate control. The whole risk management process needs to be made visible by adopting sound communication mechanisms.

'Murphy's law' states that 'anything that can go wrong, will go wrong'. Undoubtedly risks will occur in your project. Most of the risks that you will experience have already been identified in previous projects. Learn the lessons from those. To ensure that those risks have the least impact, you need to adopt structured risk management within your overall project planning.

SAMPLE QUESTIONS

1. Which of the following best defines a contingency action?
(a) An activity that is planned to take place if a risk materialises
(b) An action taken at the start of a project which reduces the potential damage if a certain risk does materialise
(c) An agreement that the users accept a particular risk
(d) An activity which prevents a risk from materialising

2. What is maintenance of a risk log designed to do?
(a) eliminate risk
(b) save money
(c) control risk
(d) prevent system development failure

3. Which of the lists below best identifies what is examined when a risk is assessed?
(a) Probability, proximity, owner
(b) Impact, probability, team experience
(c) Cost, benefit, business case
(d) Probability, proximity, impact

4. Which of the following is NOT an action that would mitigate a risk?
(a) accept it
(b) prevent it
(c) reduce it
(d) transfer it

ANSWERS TO SAMPLE QUESTIONS

1. (a) 2. (c) 3. (d) 4. (a)

POINTERS FOR ACTIVITIES

ACTIVITY 7.1

Among the facts that could lead to difficulties in the Canal Dreams scenario are the following:

- The move to 24/7 ebookings will put a much heavier stress on the Canal Dreams IT infrastructure;

- Secure money transactions will need to be carried out over the internet with customers;

- Customers will access the Canal Dreams ebookings system directly;

- New qualified staff need to be recruited for the new network support centre;

- Existing telesales staff at the bookings call centre will become increasingly redundant as ebookings pick up – or may need to be retained if they do not.

ACTIVITY 7.2

For example, the following risks might be identified. Other risks can certainly be added.

Issue	Possible risk
The move to 24/7 ebookings will put a much heavier stress on the Canal Dreams IT infrastructure	1. Poor availability of systems leads to loss of possible sales 2. Poor response times lead to abandoned transactions
Secure money transactions will need to be carried out over the internet with customers	3. Customers' financial details could be hacked
Customers will access the Canal Dreams ebookings system directly	4. Poor interface leads to abandoned transactions
New qualified staff need to be recruited for the new network support centre	5. Non-availability of suitable staff leads to delays in implementation
Existing telesales staff at the bookings call centre will become increasingly redundant as ebookings pick up – or may need to be retained if they do not	6. Staff leave before changeover because of uncertainty 7. Staff demotivated in run-up to redundancy

ACTIVITY 7.3

Note that the actions below are just suggestions. You may be able to find other, perhaps more effective, mitigating actions.

Issue	Mitigating actions
1. Poor availability of systems leads to loss of possible sales	• Facility to switch to standby systems in case of failure
2. Poor response times lead to abandoned transactions	• Acquire additional hardware capacity • Conduct volume tests
3. Customers' financial details could be hacked	• Employ specialists to assess robustness against external attack • Implement recommendations of specialists
4. Poor interface leads to abandoned transactions	• Carry out usability testing and enhancement
5. Non-availability of suitable staff leads to delays in implementation	• Employ short term contract staff on attractive terms to cover short falls • Outsource management of the IT support centre to specialist company
6. Staff leave before changeover because of uncertainty	• Generous redundancy terms conditional on staying until new system is in place
7. Staff demotivated in run-up to redundancy	• Commit to retraining staff for new roles e.g. network support

8   PROJECT ORGANISATION

LEARNING OUTCOMES

When you have completed this chapter, you should be able to demonstrate an understanding of the following:

- *relationship between programmes and projects;*
- *identification of stakeholders and their concerns;*
- *the project sponsor;*
- *establishment of the project authority (e.g. project board, steering committee, etc.);*
- *membership of project board/steering committee;*
- *roles and responsibilities of the project manager, stage manager and team leader;*
- *desirable characteristics of a project manager;*
- *role of the project support office;*
- *the project team and matrix management;*
- *reporting structures and responsibilities;*
- *management styles and communication;*
- *team building and dynamics.*

8.1 INTRODUCTION

This book endeavours to describe what could be called 'best practice' in project management. Unfortunately, not all organisations follow these good practices, which is one of the reasons why there are so many project failures. In this chapter, we describe the elements of the project management structure that should exist in an organisation that is planning and executing a project. In many cases, these roles will be known by different names to the ones we use. Many of the terms we use are loosely based on those used by PRINCE2, the project management standard sponsored by the UK government. Note that we will use capitalised initial letters, for example Senior User, where this refers to a very specific PRINCE2 concept.

8.2 PROGRAMMES AND PROJECTS

A **programme** is a collection or group of related projects. IT projects were often treated as individual and separate undertakings with their own distinct aims and objectives. However it became clear that grouping projects into programmes has advantages:

- **Reduction of duplication** – two projects may be developing a similar product.

- **Co-ordination of resources** – projects inevitably place demands on an organisation's resources and it is far better to co-ordinate than compete for them.

- **Management of interdependencies** – if projects are dependent on one another they have to be properly co-ordinated. In such circumstances, a change in one project may introduce a delay or a consequential change in another.

- **Common objectives** – sometimes several different projects need to be completed for a particular business objective to be achieved.

Typically each project is assigned a project manager, while a programme is led by a **programme director** who ideally should be an influential member of the organisation's senior management team and will act as a champion for the programme. The senior status needed for this role means that it is usually only part-time. The role of **programme manager** is likely to be full-time and deals with the day-to-day co-ordination of the programme.

In the Canal Dreams ebooking enhancement project, the booking application may be one of a portfolio of projects. Canal Dreams may wish to expand and diversify their business by extending the types of holidays that they have on offer to include, for example, rented cottages or boating holidays abroad. This would require ebooking transactions with slightly different functionalities to be developed. A co-ordinated approach to implementing these varying requirements would seem to need some kind of a programme to be initiated.

8.3 IDENTIFYING STAKEHOLDERS AND THEIR CONCERNS

A **stakeholder** is defined as anyone with a valid interest in an IT project or the products delivered by it. This group includes:

- all project personnel including managers, designers and developers;

- users, sponsors and other members of the organisation affected by the project;

- suppliers of software, hardware, consultancy, etc. to the project;

- contractors and subcontractors;

- members of the business community and, of course, financial backers.

Part of the project initiation process is to establish who these people are and to identify their needs and concerns. Processes will have to be set up to ensure that their interests are represented and that they are consulted and kept informed.

ACTIVITY 8.1

List the main stakeholders in the Canal Dreams ebooking enhancement project.

8.4 THE ORGANISATIONAL FRAMEWORK

A formal management structure is needed with defined project roles:

- Named personnel should be allocated to the roles – but several people may share a role and a single person may have more than one role.
- Roles must carry appropriate authority and responsibility.
- Individuals must carry out their roles correctly and willingly and must clearly understand their objectives.
- Status within the organisation is not sufficient qualification for a particular role. Previous experience and/or training in the role is needed.

At the top of the project organisation is a project sponsor. This person would be a senior person within the organisation who would be able to champion the project at an appropriately high and influential level – normally board level. Crucially, the sponsor controls the funds to pay for the project. Below the project sponsor is the following structure:

- project board (or steering committee or project management board);
- project manager;
- stage manager;
- project team leader;
- team member.

In parallel to this structure are three other functions:

- project assurance team;
- project support office;
- configuration management team.

Most of these titles are based on the PRINCE2 terminology, so whether or not they exist within a particular project organisation will depend on the extent to which the organisation embraces the PRINCE2 approach. However, the roles that they represent should be present in all projects to a greater or lesser degree, although they may have different names. Where a project is small, for example, the project manager, stage manager and project team leader roles could be carried out by a single person.

Having established who the project sponsor is, it is necessary to establish a body to control the project – the **project authority**. The British Standards Institution's Guide to project management (BS 6079) states that the authority for a project lies with the project sponsor. In PRINCE2, the Project Board performs this role. Project boards become more important with larger projects that have many influential stakeholders. It must be made clear who has the ultimate responsibility for the project – inevitably, this will be whoever holds the purse-strings.

8.4.1 Project board

The **Project Board** provides a forum in which critical issues can be discussed and decisions taken which are outside the remit and competence of the project manager. If PRINCE2 is not being followed, the same function may be served by a **steering committee** or a **project management board**. This group should include the project sponsor or their representative. In PRINCE2 terms this person is known as the **Executive** and is joined by a **Senior User** and a **Senior Supplier**, although in certain circumstances the Executive and the Senior User could be the same person.

Members of the board must be able to make decisions for their areas of responsibility without having to refer to higher authority. If they are at too low a level they will constantly be referring decisions back to their managers, which can delay the project. It can also be a problem if they are at too high a level, as important and busy staff may attend meetings infrequently, leading to ineffective decision-making.

All projects should derive from the organisation's business strategy and be designed to meet specific business and corporate objectives. It is therefore the role of the Executive, apart from looking after the money, to ensure that the project:

- stays in line with the corporate objectives;

- meets the business requirements;

- retains its business case (that is, that the benefits continue to outweigh the costs of the project).

The role of the Senior User is to safeguard the interests of the users and make sure that their requirements are met. He or she will be responsible for ensuring that the user requirements have been captured and deliverables have been signed off as acceptable. With the best will in the world, requirements will change over time. The Senior User must therefore ensure changes to requirements are in line with the business needs, can be justified and do not jeopardise the project as a whole. As was seen in Chapter 4, perfectly valid changes can delay the project to such an extent that it will not deliver the expected benefits.

The role of the Senior Supplier includes ensuring that adequate technical resources, in terms of both skilled staff and the software and equipment that they need, are available to the project. They should also support the development team and, when necessary, represent their interests as difficulties arise, for example in the face of continual change requests.

If all or part of the project is contracted to an external supplier, the Senior Supplier should come from that organisation. If there are a number of contractors then more than one supplier representative may be required. However, this could detract from the main purpose of the Project Board and therefore it may be more effective to set up a separate group to manage the external suppliers.

The **quality assurance function** could also have a representative on the board or may report via another member of the board, usually the Executive.

If the project is large, it may be necessary to have several user representatives. Similarly, if it is geographically dispersed then additional representatives may be required. Care must be exercised to ensure that the board does not get too big and thus become ineffective. Subgroups may have to be set up to give a voice to all those who should have one, without detracting from the efficient working of the board. This is always a balancing act, but smaller groups are usually more effective when it comes to decision-making.

The infrastructure management of the business, responsible for the platforms on which the delivered IT system will run, may also be represented. If the development staff have a different manager to the infrastructure support staff there could be a conflict of interest between the two. However, infrastructure management is often represented through the Senior Supplier.

The Project Board is a decision-making body. It holds the purse-strings through the Executive and therefore has ultimate control of the project. The project manager would normally be appointed by and report to the board. The board establishes the terms of reference and provides the management framework within which the project manager operates. It is the final arbiter on whether the project has met its objectives. In summary, the board initiates the project, controls its execution and eventually closes it down. It approves the following:

- project terms of reference (including the project initiation document);
- business case;
- budget;
- project plans;
- changes to project plans;
- quality plans, control and assurance processes;
- risk assessment and contingency plans;
- major changes to project requirements.

It receives feedback on the progress of the project from the project manager and also from the quality assurance function. The feedback enables them to sign off each stage of the project or other activity as defined in the plan, or require that it be reworked. It thus exercises overall control of the project.

8.4.2 Project manager
The project manager is pivotal in the organisation of the project and has overall responsibility for the day-to-day management of the project. The role of the project

manager is to ensure that the project is delivered on time, within budget and to the specified quality. A daunting task!

The project manager produces, with the help of others, the various plans for the project (see Chapter 2). Monitoring progress against the plan and making adjustments are ongoing for the duration of the project. As milestones are reached, progress will be reported to the board and team members (see Chapter 3). The project manager is set tolerances for activity completion and costs within which to work. Any deviation in the plans likely to take the project outside these should be reported to the board with recommendations about the actions the project manager feels are necessary to correct the situation or to mitigate its effect.

Inevitably, changes come along; the project manager assesses their impact on the project and reports to the board. It is the responsibility of the board to decide whether or not changes should be implemented although, as seen in Chapter 4, this responsibility may be delegated, within constraints, to a change control board.

Should a risk materialise, the project manager will approach the board for permission to put into effect the agreed contingency plan. Logs of both change requests and risks are kept (see Chapters 4 and 7, respectively). The project manager has the responsibility of making sure that they are kept up to date.

Because project managers have to lead their projects, they must be able to set clear goals so that those who report to them are aware of what is required. The simple production of plans and schedules does not by itself achieve this.

8.4.3 Stage manager

In larger projects, someone may be appointed to the position of stage manager. This role is similar to that of the project manager but is specific to the particular stage. Detailed plans, work schedules and monitoring then become the responsibility of the stage manager rather than the project manager. This does not mean that the project manager gives up all responsibilities; more that much of the routine work is carried out by someone else, who may have particular technical expertise in dealing with a particular stage. The project manager is still responsible to the board for the progress and success of the project.

The appointment of stage managers appears to be rare as team leaders are usually able to carry out these team leadership roles. The Stage Manager role has therefore been dropped from more recent versions of PRINCE2.

8.4.4 Team leader

Team leaders are close to the action. A team leader may be responsible for a specialist group of analysts, designers or programmers and hence work on a very specific part of the life cycle. Alternatively, a team leader may lead a mixed team, in which case the responsibilities would encompass the whole of the life cycle. A team leader usually needs technical experience and knowledge. One very specialised area is that of testing, where a team becomes expert at the creation of test data based on the requirements specification, which is then run through the software to see whether or not they meet those requirements.

Team leaders have the task of allocating authorised work packages to specific individuals and helping them to complete the activities within the scheduled time scales. They also act as mentors or advisers when necessary. They will be most aware of how well a particular part of the project is going. They should be able to quickly notify the project manager of delays so that action can be started to bring the project back on course before it goes seriously adrift.

8.4.5 Team member

At the bottom of the pile is the team member. However good the structure may be, without competent team members the project will not succeed. Ideally, project managers should be able to select their teams. This is rarely possible, so the project manager has to be familiar with the qualities of the staff allocated to them and assign them to the tasks which most match their capabilities. In larger organisations in which team leaders or stage managers exist, they may undertake this task or share it with the project manager.

As we will see in the section on matrix management, specialist staff often have only a short-term project relationship with the team leader, while their longer term development and deployment is the responsibility of a separate technical head.

8.4.6 Project assurance team

This is usually a function rather than an actual team. The project board could fulfil the role. However, depending on the size and nature of the project, it is normally delegated to one person or a small group. The tasks are the same; it is just the degree of activity that differs. Members of the project assurance function are appointed by and report directly to the project board, not to the project manager. Their role is primarily quality assurance (see Chapter 5). They are responsible for checking that the quality control activities in the quality plan are carried out, standards are observed and procedures are followed. Being independent of the project manager, they can give an objective view on the quality of the products delivered and not just the timeliness of their arrival.

In the early stages of a project, the team would contribute to the setting of standards, the creation of procedures, the establishment of the quality review and quality assurance processes and the quality plan. This is particularly useful if the project is venturing into areas of technology that are new to the organisation.

A large, complex IT project may need several people for this activity, each with their own specialism, such as security. Three specific roles sometimes identified are:

- **business assurance co-ordinator**, who would, for example, make sure that the IT application under development is compatible with existing business procedures;

- **technical assurance co-ordinator**, who would ensure that the operational environment is not compromised by the new system;

- **user assurance co-ordinator**, who ensures the application meets user needs.

ACTIVITY 8.2

In the Canal Dreams ebooking enhancement project, the main driving force behind the new internet-based booking system is the managing director. He has given you a contract to manage the project. You are in regular contact with the manager of the telesales call centre, local boatyard managers and the directors of marketing and of finance. You also need to communicate regularly with the IT infrastructure manager at Canal Dreams. The software is to be supplied by the XYZ software house and you have had several meetings with one of their account managers. Identify which type of role on the Project Board would be likely to represent each of the various stakeholder groups in the scenario.

8.5 DESIRABLE CHARACTERISTICS OF A PROJECT MANAGER

The role of the project manager is pivotal. IT project managers have traditionally progressed from technical areas such as software development through system design and team leadership to project management. This, however, may be changing, as the project management role nowadays often involves managing the external suppliers of services. Thus, while it is useful to have a good technical background this is not always essential in order to manage an IT project.

It is more important to have other skills and characteristics. A key quality is good communication skills at all levels. The project manager may need to present to higher management a convincing case for a course of action. Stakeholders have to be kept informed and the project teams must be motivated. If project managers are to gain the respect and confidence of those around them, they have to be effective communicators, good managers and effective organisers.

They need to have skills in:

- leadership;
- motivation;
- planning;
- negotiation (being firm, flexible, and able to compromise where appropriate);
- delegation.

They need to be:

- responsible;
- reliable;
- available (not just for this project but contactable at all reasonable times);
- intelligent;
- sociable (able to mix well);

- approachable (they should be good listeners);
- knowledgeable in the business area for the particular project.

These characteristics do not just arrive with seniority. A potential project manager must possess some of these attributes and will need development in deficient areas before becoming fully effective.

Although this list may seem daunting, the ability of people who seem quite 'ordinary' to become competent project managers through application, self-discipline and appropriate training is remarkable.

8.6 PROJECT SUPPORT OFFICE

The qualities we have associated with project leadership seem heroic, but projects have many mundane clerical activities which have to be performed regularly, effectively and efficiently. The project manager is accountable for them, but given the pressures on project managers, it is more effective to delegate these tasks. This gives rise to what has become known as the **project support office** (PSO).

The organisational structure that has been described above is usually set up for a specific project. This means that the project board and project manager are appointed for a single project. The project support office, on the other hand, may be to a greater or lesser extent a permanent organisation which supports several often inter-related projects. In these cases it is convenient to combine project support with programme support – hence we have a programme and project support office (PPSO) or simply a programme management office (PMO).

The precise tasks this office undertakes will vary, but the following list is typical:

- time recording;
- updating of project plans;
- risk log maintenance;
- issues (change) log maintenance;
- arranging meetings;
- issuing agendas;
- taking minutes;
- chasing actions;
- configuration management.

8.6.1 Time recording
Time recording is essential for project control. Clerical staff can collect and process the information required. The team leaders or project managers should be aware of what is happening but, relieved of the routine work, they can focus on checking that effort expended is consistent with that expected and, if not, take the necessary action.

8.6.2 Updating plans

As details of progress are passed to the PSO, staff can update the plans and highlight problems to the team leader or project manager. In the event of changes to project team membership, the PSO can revise the plans so that the project manager is aware of the impact of such changes and can make any modifications necessary or alert the board to potential problems.

8.6.3 Maintenance of logs

The PSO can maintain project logs. The project manager can be warned of approaching risks via the risk register (see Chapter 7). Risks that have passed can also be noted. Requests for change (RFC) are recorded as they arrive and then passed on for action (see Chapter 4).

8.6.4 Arranging meetings

A great deal of time, most of which involves finding dates and times at which all parties are available, can be spent on arranging meetings. Aids such as electronic diaries make this easier, but it can still be time-consuming. Delegation to the PSO is natural. Having agreed dates and times, the PSO can:

- organise the necessary room and other requirements such as catering;

- issue the agenda – most meetings have a set agenda and it is simple to check with the chair of the meeting to see if any changes are required;

- circulate other documents needed – the PSO may already carry out the configuration control function which controls the versions of documents;

- record and distribute the minutes of meetings – the key thing is to ensure that all actions are identified, along with who is responsible for carrying them out and in what time frame.

8.6.5 Configuration management

The whole of Chapter 4 has been given over to change control and configuration management. The ability to keep track of documents and products to ensure that everybody is working with the latest version is very important to the success of the project. Like the PSO, within which it may function, configuration management has a life beyond the duration of the project, as the delivered IT system will continue to require amending and updating until it is finally replaced.

8.7 PROJECT TEAM

'A project team is defined as a small number of people with complementary skills who are committed to a common purpose, performance, goals and approach who are directly or indirectly accountable to the project manager.'

Michael W. Newell and Marina N. Grashina, *The Project Management Question and Answer Book*, American Management Association, 2004.

Project teams work in a different way from operational staff. A project team is brought together for the sole purpose of achieving the project objectives. On completion of the project, the team is disbanded. Team members may be drawn from a **specialist division** within the organisation, such as the IT department, to work on the project and may be assigned to other projects when it is over. Others may be drawn from **functional departments** and will go back to their original jobs at the end of the project. The impact of being in a project team will be quite different for the two types of people, as will their expectations during and after the project.

Project team members are not always located together, although this is best. If the team is drawn from different departments, they could be located in different parts of the country. It may be possible to bring them together for short periods. However, they could be located in different countries, which would make that much more difficult. Part of the solution to these problems could be matrix management.

8.8 MATRIX MANAGEMENT

So far it has been assumed that all staff report to the project manager, either directly or through team leaders or stage managers. This is the best way of managing a project but is not always practical, particularly if team members are drawn from a variety of departments.

Senior staff on a project – that is, the stage managers and team leaders where they exist – should always report to the project manager. If this is not the case the exercise of managerial control may be weakened to a point where it would be impossible, with any certainty, to keep to a delivery plan.

In matrix-managed teams, an individual reports to more than one person. One example of matrix management can be found on board a naval ship. The captain of the ship has overall responsibility for everything that happens on the ship when it is at sea. Its complement of sailors will include navigators, engineers, electricians, cooks, medical staff and others. Each specialised trade will also have a shore-based manager to whom they will report and who is responsible for their performance on the ship and their training when ashore.

In any organisation, it is probable that a number of projects would be under way at the same time, each requiring a range of different skills. For example, business analysts may help to elicit requirements from the users and ensure that they are clearly understood by the technical members of the project team. Analysts and designers would then be required to convert the requirements into a design specification. They could well be part of a specialist department within the organisation and report to a separate manager. There is likely to be another department which manages the IT infrastructure. They could second someone to the team to look after their interests and guide the designers so that the proposed solutions are compatible with the existing structures.

The developers (or programmers) may be part of a separate pool of development specialists, or even come from a specialist group outside the organisation. The testing team may come under a different department with its own line management. Putting all these together gives rise to a matrix structure as shown in Table 8.1.

Table 8.1 Example of a matrix organisation

Project	User department	Business analysis department	Infrastructure management	Software development department	Software assurance
A	X	X		X	X
B	X		X		
C			X	X	X

This sort of structure has the advantage that the project manager has a team of specialists and can concentrate on the project in hand and not be concerned with issues such as the long-term development of the staff involved, as this would be the responsibility of the line managers. The big disadvantage is that the line managers may call their staff off the project for some more urgent task because of an unforeseen event such as staff illness, making planning and control more difficult. These issues should be discussed at the beginning of the project and a strategy agreed by the project board, to which all line managers sign up.

Even where all staff, including the project managers, work for the same line manager, they could be allocated to more than one project. Each person may have their own perceived priorities and may put more effort into one project than another, perhaps because one task is easier or more interesting or more rewarding. This can lead to slippage on some projects, while others progress to schedule.

We noted in Section 8.2 that where there are many projects being carried out at the same time, particularly if they are related, there is often an umbrella organisation known as **programme management**. Where resources are limited, which is the normal situation, a **programme board** would have the remit to make the final decisions about the allocation of resources between projects. While this may be irksome for the individual project manager, it is to the benefit of the organisation as a whole. Obviously plans have to be revised to take account of this situation and project managers cannot be held responsible for delays to their projects due to such changes.

With matrix management, the level of control that the project manager will have over the project team will vary. Where an individual is seconded to the project for its duration then the project manager has greater control – as with the ship's crew. This is in many ways the ideal, but it does have its drawbacks. With every project, the level of resource required will vary over time. For example, in the early, analysis stages of a project, several analysts but only one part-time designer may be required. However, as the analysis phase is completed, more designers may be required, but fewer analysts. If these staff were permanently with the project they would be under-utilised at times, adding unnecessary costs to the project.

ACTIVITY 8.3

List the types of people (including external suppliers) required to implement the Canal Dreams ebooking enhancement project.

8.9 TEAM BUILDING

A team in an IT project is usually a collection of specialists with a requirement to work together towards a common goal. It is usual for the composition of the team to be based on a number of compromises; it will never offer the ideal combination of skills and personal characteristics. Making the group of individuals into an effective team is an interesting and rewarding management challenge. There are many commonsense aspects and a number of theoretical approaches to developing an effective team.

The Tuckman-Jensen model describes four basic phases through which a team goes before it becomes fully effective:

- **Forming.** The group members have just been brought together and are probably hesitant about their new environment, unsure of their new colleagues and possibly nervous about future developments. Members are polite to one another, tend to accept authority and tread carefully. Some initial contact with colleagues reveals common ground and possible allegiances.

- **Storming.** Individuals within the group have started to assert themselves and to form alliances. Conflict may arise as 'pecking orders' become established. Aims and objectives are becoming clearer but there are different views on how to proceed with the tasks ahead. Members now have a sense of belonging to a team, are gaining confidence and are likely to challenge the proposed methods of working.

- **Norming.** Internal conflicts are hopefully resolved and the team members feel more comfortable and relaxed with their colleagues and their new surroundings. An acceptance of common values and behaviours develops, with open communication and constructive cooperation. The team starts to work as it should, with its overall capability being greater than the sum of its parts.

- **Performing.** The team is fully functional and has become a cohesive unit. Team morale is high, with good cooperation between members and a shared responsibility for the common goal. Team members are working hard and getting satisfaction as the team achieves its goals.

A fifth stage, **adjourning**, is sometimes added to describe when the team breaks up at the end of the project. The Tuckman-Jensen sequence of phases clearly represents the ideal team development. Good team leadership and people management are essential to allow the team to progress through the phases, and indeed the final stage may never be achieved if the personnel or the project circumstances are not right. It is quite possible to slip back a stage or two if unexpected developments are not well managed – for example:

- changes in team personnel – new arrivals can disrupt team morale and stability;
- a change in direction of the project, which means much team effort has been wasted;
- a change of leadership, which may mean the team needs to adapt to a new management style and could revert to the storming stage.

8.10 TEAM DYNAMICS

Building a team involves finding people with the appropriate skills who are available when you need them and are motivated to perform the tasks required of them. However, if the team are to work well together, a satisfactory mix of personality types and personal attributes is essential. A system for analysing and categorising people's personal characteristics was developed by Belbin, who defined nine **team roles**:

- **Shaper** – an energetic team member with a strong need for achievement who drives the team along;
- **Plant** – a creative and innovative team member (the term 'plant' is used because it was found that planting such a person in an uninspiring team was a good way to improve its performance);
- **Resource investigator** – a team member who makes contacts outside the group to bring in ideas and information and to acquire materials/resources;
- **Co-ordinator** – a chairperson who promotes decision-making and delegates well (not necessarily the team leader);
- **Monitor evaluator** – a team member who is analytical and able to assess ideas and options but is not creative;
- **Team worker** – a team member who helps to maintain team spirit and cohesion;
- **Completer finisher** – a conscientious and painstaking team member who is concerned with getting things finished (this is a very important team trait);
- **Implementer** – a team member who attends to details, is hard-working and organises the practical side of the project;
- **Technical specialist** – someone who can provide the team with technical expertise.

COMPLEMENTARY READING

For further details, see R.M. Belbin, *Management Teams: why they succeed or fail*, 2nd Edition, Elsevier, 2003

It is not suggested that a team has to have one person of each type or that each team member falls into only one role. An individual can have attributes from a number of different roles. The idea is that a team needs a satisfactory mix of roles played by its members to perform well. A Belbin analysis may help to define a team weakness, or indeed to clarify the reasons why a team is not performing well or why conflicts keep arising in an otherwise skilled, competent team.

8.11 MANAGEMENT STYLES

There are as many styles of management as there are managers. The different styles and approaches depend as much on the personality and capability of the manager (or leader) as on the environment and the prevailing circumstances. The manager may well have a natural preference for a certain style, but will have to vary their approach depending on the circumstances in order to be successful.

For example, a **task-orientated** leadership style is sometimes distinguished from a **relationship-orientated** leadership style. Task orientation focuses on the technical aspects of the work, while relationship orientation emphasises such things as individual motivation and team morale. When a team is developing – the forming stage – it will need more direction than when it is fully functioning – the performing phase. A task-orientated approach would be more effective than a relationship-orientated approach because the team members are not yet familiar enough with the tasks to be carried out and the environment in which they are to be carried out.

This distinction between task orientation and relationship orientation overlaps with that between **autocratic** and **democratic** styles of management. When staff are new to a project a more autocratic approach may be called for, as staff are not yet sufficiently knowledgeable to be involved in complex decision-making. Individual team members will react differently to these two approaches. Some people prefer to have clear direction and to leave decision-making to others, whereas other team members like to feel they have a part to play in establishing the direction of the project and wish to provide an input to the decision-making process. An intermediate approach is described as **consultative**.

- Autocratic management provides clear direction and quick decisions. The leader is seen to be decisive, firm and effective. However, it can be synonymous with an uncaring, remote, unapproachable, controlling or bullying management style. An autocratic leader can demotivate a team that is working in a climate of fear (members may seek an alternative project).

- Democratic management shares responsibility for decision-making and for the team's performance: thus the team is likely to be more committed to the project. This style can be perceived as weak and management can be seen as avoiding responsibility for difficult decisions. It may be difficult to enforce discipline if conflicts develop.

- Consultative management seeks the opinions and views of the team prior to a decision being made. Although team members are consulted, the final course of action may or may not reflect their views. Consultative management is seen as a compromise between autocratic and democratic management, with the benefits, and possibly the pitfalls, of both.

A politically acceptable and effective manager would need to stay away from extremes and be able to adapt styles of management to the situation and the people involved. Ideally such a manager could be decisive and effective where a quick decision was required and approachable and consultative on other occasions, and thus gain commitment from the team where joint decisions and consensus were appropriate.

Whichever approach is used, the project manager must be an accomplished communicator and have the ability to use the most appropriate **methods of communication** in order to be fully effective.

8.12 COMMUNICATION METHODS

Methods of communication for IT projects can include:

- memos;
- newsletters;
- meetings;
- noticeboards;
- presentations;
- progress reports;
- telephone conversations;
- text messages;
- email messages;
- letters;
- drawings;
- one-to-one conversations;
- video conferencing;
- intranets and extranets;
- social networks.

Methods can be categorised as **active** or **passive** and as **formal** or **informal**. **Active** methods, such as a telephone conversation, require a response or reaction so that there is reinforcement or confirmation that the information or message has been received and understood. **Passive** methods such as a newsletter require no such confirmation. They leave to chance whether anyone reads or understands the material and should not be used for important messages.

Formal methods of communication have a set structure, such as a meeting of a project board, in contrast to **informal** methods such as a conversation, which carries no particular format and is not usually recorded.

It is also possible to categorise methods depending on whether the participants have to be available at the same time and/or in the same place for communication to take place.

ACTIVITY 8.4

Categorise each of the methods listed above as same time/same place, same time/different place, different time/same place, or different time/different place.

By categorising methods in this way it is possible to assess the suitability of a particular method. It is best to put together a **communications plan** setting out the methods of communication that should be used during the project for each set of circumstances and exactly how they should be applied. For example, you may decide that a short weekly face-to-face meeting (an active, formal method) with the project sponsor would be better than an email.

By setting out a detailed communications plan with all stakeholders, rather than leaving things to chance, you stand a much better chance of getting it right. Below is an example of an entry for one of the types of communication selected for a project:

Name/description	Joint application development session
Target audience	User representatives and business analysts
Purpose	To elicit user requirements
Frequency/event	23rd April
Method of communication	Away day
Responsibility	Ann Smith (Project Manager)

8.13 CONCLUSION

This chapter has only been able to touch briefly upon some aspects of organisational behaviour as it relates to projects. A good book on the important issues of organisational behaviour that goes beyond the immediate demands of the BCS Foundation Certificate in Project Management for Information Systems is:

COMPLEMENTARY READING

John Arnold, Cary Cooper, and Ivan Robertson, *Work Psychology: Understanding human behaviour in the workplace* (4th Edition), FT Prentice-Hall, 2004

SAMPLE QUESTIONS

1. Which of the following is not a name given to the group which has responsibility for committing resources to the project and approving variations to the project's objectives?
(a) project board
(b) project management board
(c) project support office
(d) steering committee

2. Which of the following terms is used to describe an organisational structure where staff are responsible to a project manager for the duration of a project but also have a manager who is responsible for their long-term staff development and work programme?
(a) the Tuckman-Jensen model
(b) configuration management
(c) matrix management
(d) task orientation/relationship orientation

3. At which stage of team formation does a team become fully functional as a cohesive group?
(a) performing
(b) norming
(c) storming
(d) forming

4. Which of the following is an example of different time/different place communication?
(a) a project board meeting
(b) a checkpoint report
(c) a telephone conversation discussing a problem a user has with an IT system
(d) video conferencing with the supplier of a software application

ANSWERS TO SAMPLE QUESTIONS

1. (c) 2. (c) 3. (a) 4. (b)

POINTERS FOR ACTIVITIES

ACTIVITY 8.1

Stakeholders in the Canal Dreams project include existing IT support staff, customers, boatyard staff (who will be checking in customers at the start of their canal holiday), current telesales booking staff, the managing director, the marketing director, the central finance department, the personnel department, the premises manager (who has to find room at head office for a network support centre), IT equipment suppliers and software suppliers.

ACTIVITY 8.2

The Senior User represents the interests of boatyard staff, telesales booking staff and the marketing and central finance department.

The Senior Supplier reflects the views of existing IT support staff, IT equipment suppliers and software suppliers.

The Executive represents the interests of the managing director and the marketing director (who may be the member of the senior management team with the strongest interest in the project).

It could be argued that the personnel department, who will have to recruit new network support staff and perhaps relocate existing staff as a result of this project, and the premises manager, who will have to find accommodation for the new network support centre, are suppliers of services to the project.

Who will represent the interests of the customers of Canal Dreams? It may be the marketing department, who may have conducted market research into the preferences of potential customers, or the booking and boatyard staff, who have had most contact with customers in the past. These would be represented on the project board by the Senior User.

Note that terms like 'Senior User' simply represent roles, and more than one person might undertake that role. Each project must form a project board with representation that is sensible for that project.

ACTIVITY 8.3

- Users;
- Analysts;
- Designers;
- Database designers;
- Network/telecommunications specialists;
- Hardware specialists;
- Software developers;
- Testers;
- Trainers.

ACTIVITY 8.4

	Same place	Different place
same time	meetings presentations (normally) one-to-one conversations	telephone conversations video conferencing
different time	noticeboard	memos newsletters progress reports text messages email messages letters drawings intranets and extranets social networks

INDEX